Information Systems in Develo

Information Systems in Developing Countries

Theory and Practice

Edited by

Robert M. DAVISON
Roger W. HARRIS
Sajda QURESHI
Douglas R. VOGEL
Gert-Jan de VREEDE

City University of Hong Kong Press

First published 2005
Printed in Hong Kong

ISBN: 962-937-110-3

Published by:
City University of Hong Kong Press
Tat Chee Avenue, Kowloon, Hong Kong
Website: www.cityu.edu.hk/upress
E-mail: upress@cityu.edu.hk

Contents at a Glance

Detailed Chapter Contents

3.   Context-Specific Rationality in Information Systems

4.   Support Environments for E-Learning in Developing Countries

Section II— Telecentres

5. Explaining the Success of Rural Asian Telecentres

Roger W. Harris

10.   Online Success in a Relationship-Based Economy—
Profiles of E-Commerce in China

11.   E-Procurement by the Brazilian Government

Preface

In contemplating and planning for this volume, a major influence was the ever increasing awareness of the digital divide—the yawning gap between the information and communication technology (ICT) haves and have-nots. Despite this awareness, and despite a number of notable success stories, much of the developing world was still lagging far behind the developed world five years ago. To be sure there were barrels of rhetoric, oceans of urgent commentary and even a modicum of good will. But what real difference was it going to make to the lives and futures of those, through accident of birth or fate, who were born in the less developed countries of the world? A major objective of this volume is to raise awareness further, indeed to raise the consciousness of those in power, those with money to spend (or not to spend) altruistically or for profit, to the ICT plight of the majority of the world's population. Allied to this objective is the desire to raise an equal awareness of what can be done, often for surprisingly little outlay of funds, though the demands on time and energy may be considerable. In this respect, many of the chapters in this volume document cases of ICT projects in developing countries that are achieving remarkable success despite constricting circumstances.

A primary source of material for this book comes from what we consider to be "best papers" published over the last five years in the *Electronic Journal of Information Systems in Developing Countries* (EJISDC)—http://www.ejisdc.org. We selected these best papers for their power to convey a message appropriate to the book's overall theme, in addition inviting complementary chapters from established authors in the field. The resulting compilation is organized in four sections of roughly equal weight. Each chapter can be read as a separate entity, but readers will find that there is considerable commonality of purpose across the fifteen chapters in the book.

Our intention is that readers should be both informed and challenged by these chapters. Much of the material will be intuitive to those who are already working in this domain, though the precise details of the cases may be new. For others, this may be the first time that they have come across such material—and it may be something of a shock. A culture shock no less that may threaten the validity of much that they have held to be true until now. To these readers, we seek your open-mindedness and willingness to reflect on the realities of life in societies that are relatively impoverished with respect to ICT availability and use. We hope that all readers will be inspired by the possibilities and opportunities that these chapters describe, and that they will be encouraged to go out and make a difference, however small, in their own societies and lives.

A preface serves not only as a place to write a rationale, but also as a place to write acknowledgements. We are indebted first to our authors for their endeavours in promoting the cause of ICTs in developing countries. We also gratefully acknowledge the support of the university-based sponsors who have funded the EJISDC from which many of the chapters in this book were drawn. These include the City University of Hong Kong, the University of Malaysia Sarawak, Erasmus University, Delft University of Technology, and the University of Nebraska at Omaha.

Robert Davison, Doug Vogel, Roger Harris, Hong Kong
Gert-Jan de Vreede, Sajda Qureshi, Omaha, Nebraska, USA
May 2005

List of Illustrations

Figures

Tables

Acronyms and Abbreviations

Acronyms	Names in Full	Appear First on Page
FAVC	Fiji Audio Visual Commission	241
FINTEL	Fiji International Telecommunications Limited	240
FOOD	Foundation of Occupational Development	87
FRIENDS	Fast, Reliable, Instant, Efficient, Network for Disbursement of Services	276
FTIB	Fiji Islands Trade and Investment Bureau	239
G2B	Government-to-Business	59
G2C	Government-to-Citizens	59
GATS	General Agreement on Trade in Services	230
GIS	Geographical Information Systems	9
GPS	Global Positioning System	9
IAPs	Information Access Points	118
ICTs	Information and Communication Technologies	1
IDRC	International Research Development Centre	86
IEC	Information, Education and Communication	114
IICD	International Institute for Communication and Development	120
IISc	Indian Institute of Science	257
IITM-K	Indian Institute of Information Technology and Management—Kerala	278
IITs	Indian Institutes of Technology	257
IS	Information Systems	1
ISP	Internet Service Providers	224
ISPAI	Internet Service Providers Association of India	264
IT	Information Technology	148
ITEBS	IT Enabled Business Services	224
ITES	IT-Enabled Services	229
ITU	International Telecommunications Union	2
KELTRON	Kerala State Electronics Development Corporation	279
KIK	Keltron Information Kiosks	279
LAN	Local Area Network	279
LNF	Large National Firms	264
MCA	Master of Computer Applications	283

Acronyms	Names in Full	Appear First on Page
MIT	Massachusetts Institute of Technology	221
MITech	Masters Programme in Information Technology	278
MMORG	Massively Multiplayer Online Role-Playing Games	185
MMS	Multimedia Messaging Services	185
MNE	Multinational Enterprises	264
MPAS	Ministry of Social Security	194
MSSRF	MS Swaminathan Research Foundation	87
MTNL	Mahanagar Telephone Nigam Limited	256
NASSCOM	National Association of Software and Services Companies	285
NBA	National Basketball Association	184
NCST	National Centre for Software Technology	257
NGOs	Non-Governmental Organizations	89
NICNET	National Informatics Centre Network	257
NICTS	National Information and Communication Technology Strategy	132
NTP	National Telecommunication Policy	256
OECD	Organization for Economic Co-operation and Development	104
OFC	Optical Fibre Cable	279
OLE	Online Learning Environment	62
OSI	Open Society Institute	89
PANTLEG	PANAsia Telecentre Learning and Evaluation Group	86
PBG	Performance Bank Guarantee	259
PEARL	Package for Effective Administration of Registration Laws	277
PICs	Public Internet Centres	89
PPGIS	Public Participatory Geographical Information Systems	19
RICs	Regional Information Centres	131
RITI	Regional IT Institute	70
S@T	Somos@Telecentros Network	121
SAAFI-Agro	Decision Support System for Financial Evaluation of Agroindustrial Projects	213
SARI	Sustainable Access in Rural India	119

Acronyms	Names in Full	Appear First on Page
SARS	Severe Acute Respiratory Syndrome	187
SDE-CAF	Strategy, Development, Evaluation, Content Management, Assessment, Feedback	67
SHGs	Self-Help Groups	123
Sify	Satyam Infoway	261
SII	State Information Infrastructure	273
SMEs	Small and Medium Sized Enterprises	184
SMS	Short Messaging Services	184
STS	Sociotechnical Systems	33
TANUVAS	Tamil Nadu University of Veterinary and Animal Sciences	123
TEK	Traditional Environmental Knowledge	10
TKM	Total Knowledge Management	72
TRAI	Telecommunications Regulatory Authority of India	256
TREE	Tropical Forest Ecology Experiment	17
UMIST	University of Manchester Institute of Science and Technology	78
UNCTAD	United Nations Conference on Trade and Development	230
UNDP	United Nations Development Programme	90
UNESCO	United Nations Education, Scientific and Cultural Organization	114
UNITAR	Universiti Tun Abdul Razak	70
UPS	Uninterrupted Power Supply	136
UUNet	UNIX-to UNIX Network	258
USAID	US Agency for International Development	117
VLEs	Virtual Learning Environments	63
VOISS	Virtual Online Instructural Online System	70
VSAT	Very Small Aperture Terminal	87
VSNL	Videsh Sanchar Nigam Limited	256
WAN	Wide-Area Networks	257
WTO	World Trade Organization	187

Information Systems
in
Developing Countries

Introduction

The Birth of a Journal

In January 1998, the co-editors of this book met at the Hawaii International Conference on System Sciences in preparation for a session of the conference that would be devoted to Information Systems (IS) in developing countries. In thinking about how excellent papers might move beyond the conference, we realized that there was little in the way of choice. At that time, no journal was issuing papers on IS in developing countries with any regularity. Moreover, most mainstream journals displayed little interest in such papers, given that their readers almost exclusively came from the developed countries of the world. We recognized the niche for a journal devoted exclusively to this domain. In particular we envisioned the potential for an electronic journal, which would be available to readers for free. This would allow researchers in developing countries to access and publish relevant research findings, in a timely and cheap fashion. The journal would set itself the target of disseminating best practice in the field of IS in Developing Countries to a global audience.

The Electronic Journal of Information Systems in Developing Countries (EJISDC) was conceived in 1999 and the first papers were issued in January 2000 (see www.ejisdc.org). Since 2000, we have published over 120 articles with authors from over 30 countries and over 40 countries described in papers. These articles are regularly cited in UN and World Bank reports. Readers can be traced to over 100 countries. The journal receives a healthy number of high-quality submissions, indicating that it is becoming established as a journal of some repute in its chosen domain. The journal also partners with various conferences so as to produce theme-focused special issues, a trend that we expect to continue. We intend that the journal will continue to publish articles that are both rigorous and relevant, that both inform our knowledge about and advance the theory and practice of IS in Developing Countries.

An Inevitable Luxury?

It has always been a concern of the editors that the digital divide in Information and Communication Technologies (ICTs) that separates the more developed countries of the world from the less developed would ensure that these same ICTs would be beyond the reach (financial, education, practical, technical) of the vast number of people living in developing countries. Neki Frasheri, a delegate at the Bangkok conference, referred to ICTs in this context as being an "inevitable luxury"—their acquisition by developing countries is inevitable, but it will cost them dearly. In selecting the electronic domain for EJISDC, we certainly ensure that the papers we publish are only available to those (few) with access to the Internet—all papers are published purely as pdf files on the Internet. Nevertheless, this access is free of charge—a characteristic extremely seldom associated with print journals, unless they have generous sponsorship. Given that EJISDC is essentially self-funded, with no budget to speak of, a zero cost strategy is essential. In its first two years, a web server was provided at the Universiti Malaysia Sarawak. Since 2001, it has been hosted by the City University of Hong Kong.

Why Not a Book?

On the sidelines of the 15th Bled eCommerce Conference held in June 2002 in Bled, Slovenia, Roger and Robert formulated the idea of crafting an edited volume of selected papers that focus on the experience of Information Systems in developing countries. Our initial idea involved selecting some of the best papers published since the inception of the journal and asking the authors to improve them further. At the same time, we would invite selected additional authors to contribute chapters. This was therefore not a free for all—we did not issue a general purpose call for papers on international mailing lists. Some of those we invited declined the opportunity, but in the end, 15 author teams committed to writing chapters for the current volume. These 15 papers are organized into four sections, each section managed by one co-editor. The co-editors will introduce their sections later in the book, but the four sections are—theoretical background and culture; telecentres; applications; country specific studies.

Information Systems in Developing Countries—
Renewed Attention

In the last few years, we have observed that there is a renewed focus on the digital divide and the inequalities of ICT application between developed and developing countries. Global institutions such as the United Nations, the International Telecommunications Union (ITU) and the World Bank, as well as regional and national level agencies have set up task forces to investigate how ICTs can be enacted most effectively in developing countries.

On 17 May 2004 (World Telecommunications Day), the UN Secretary General, Kofi Annan, noted[1] that this year's telecommunications theme is "*ICTs—Leading the Way to Sustainable Development*". The notion of sustainable development is very much at the heart of the application of IS in developing countries—it has to be self-

---

[1]   www.un.org/News/Press/docs/2004/sgsm9294.doc.htm

sustaining if it is to thrive in the long run. This is, indeed, a theme that is taken up in many of the chapters in this book. ICTs thus have the potential to serve as essential components of a strategy to achieve economic progress, as well as social change—if they are affordable and appropriate to the local context. In Geneva, in December 2003 at the first meeting of the World Summit on the Information Society (www.itu.int/wsis), "leaders set out a shared vision of how the world can eliminate the digital divide in content and physical infrastructure". But putting this vision into practice is self-evidently quite a different matter.

In commissioning chapters for this book, we specifically asked some of the authors to describe how ICTs do make a difference, do bridge the digital divide, do approach self-sustainability and do create new opportunities. Examples of these applications are described in chapters about India, China, Fiji and Thailand.

Information Systems in Developing Countries— The Road Ahead

The challenge for disseminating information systems to developing countries, as we see it, lies very much in the action. We have seen all manner of policies, ideas and polemic about how important this is. Task forces, commissions and working parties are formed left, right and centre. But what of the action? What difference does all this make to the people on the ground? In this book, we go beyond the theoretical, the polemical and the philosophical to the practical. We hope that readers of this book will be stimulated by descriptions of how information systems do make a difference, do bring about self-sustaining changes.

References

Davison, R. M., Harris, R. W., Jones, A. N. and Vogel, D. R. (1998) Technology Leapfrogging for Development—A Panel, *Proceedings of the 5th IFIP WG 9.4 Working Conference: Implementation and Evaluation of Information Systems in Developing Countries*, Bangkok, February 18–20, 369–370.

Davison, R. M., Vogel, D. R., Harris, R. W. and Jones, N. (2000) Technology Leapfrogging in Developing Countries: An Inevitable Luxury?, *Electronic Journal of Information Systems in Developing Countries*, 1, 5, 1–10.

Theoretical Background and Culture

The purpose of this section is to explore aspects of culture within a theoretical background and context. Different perspectives are presented representing the rich nature of the topic. Each of the papers examines a topic and associated issues that exposes ways of thinking, feeling and reacting that may (or may not) align with varying circumstances. Reasons and implications are discussed in this generalized culture orientation. Prescriptions within the domain of each topic and perspective are provided when appropriate.

Closing the Chasm—
Enhancing Local Knowledge through Spatial Information Technologies
Gernot Brodnig and Viktor Mayer-Schoenberger

The authors note that recent years have seen a proliferation of initiatives to harness the potential of information and communication technologies (ICT) for the benefit of the poor and disadvantaged spurred on by the debate on the Digital Divide. At the same time, development programmes and projects are increasingly trying to supplement, if not substitute, Western science-based approaches with an emphasis on local knowledge. They feel that this encounter of modern technology with traditional cognitive systems is particularly prominent in the area of sustainable natural resources management. Their paper goes on to examine a subset of ICT—spatial information systems (including GPS and Remote Sensing), that collect, manipulate and distribute data on a variety of environmental factors, in order to inform and encourage sustainable resources management practices. They examine grassroots and participatory GIS approaches, with a particular emphasis on community mapping initiatives in South East Asia. The paper analyses how these projects have empowered local communities by helping them and their counterparts to find a common language and to build a bridge between traditional knowledge and the demands of a globalized world.

Endogenizing IT Innovation—
Preliminary Considerations for an Approach to Socio-Economic Development
Stephen S. Corea

The author undertakes to conceptualize the notion of technological innovation and societal development in compatible terms, in order to elaborate an innovation-based approach to promoting the socio-economic growth of societies—one with a central role for the use of information technologies. A behavioural notion of development, based upon the structural conditioning of individual and group behaviours, is found congruent, and useful, for understanding the dynamics of transformation by which a fundamental orientation towards practices of technological innovation can significantly raise the level of societal wealth over the long term. Insights are recruited from sociotechnical systems theory to inform a preliminary elaboration of principles by which social systems may be organized to generate learning behaviours that lead to continuous improvements and socio-economic prosperity. The main outcome of this paper's exploration is to develop the argument that technological innovation in less advanced societies should constitute a self-cultivated rather than an imported behavioural phenomenon. The author concludes that it is of greater priority for such societies to invest in cultivating the patterns and practices of behaviour that underpin the various IT based innovations of modernization, than to pursue alternative policies such as aggressive IT adoption or rapid technical change.

Context-Specific Rationality in Information Systems
Chrisanthi Avgerou

The author claims that most of the IS literature is based on the assumption that processes of technology and organizational innovation are driven by a mix of three rationalities, that is, the following three modes of reasoning—techno-scientific reasoning of engineering; managerial reasoning for effective organizing in terms of competitive positioning a business firm in a market context; and economic reasoning for optimizing economic utility for the individual and the firm. She challenges the assumption that IS practice can be understood as employing these types of universal rationality. In particular, she points out that the literature of development studies, and IS in developing countries in particular, suggests that the combination of rationalities that prevails in current IS and organizational development knowledge is ineffective in mobilizing local human capabilities to improve socio-economic conditions. Consequently, she argues for a context-specific notion of rationality for the study of information systems. Drawing from Foucault's concept of knowledge as being inseparable from power relations within society, she introduces the concept of "organizing regime" to refer to historically constructed rationalities within which IS innovation is embedded.

Support Environments for E-Learning in Developing Countries
Noorminshah Iahad and Georgios A. Dafoulas

The authors discuss the basic concept of E-Learning, its requirements and advantages in the scope of developing countries with special emphasis on corporate organizations and the higher education sector. They suggest that the successful implementation of an E-Learning environment depends on the development of the appropriate infrastructure and the support environment but note that problems such as an inadequate number of personal computers per household, poor Internet connectivity, and insufficient bandwidth need to be resolved before an appropriate support environment for E-Learning can be created. They feel that such an environment must of necessity ensure the relevance and quality of the Content, Assessment, Course Management and Communication and stress that the environment must not only consider technology-related aspects but also pedagogical aspects of E-Learning. This concludes that this will ensure that the demand for a learner-centred learning environment (that emphasizes collaboration, personalization and feedback) will be fulfilled. They go on to suggest that strong support for a learner-centred E-Learning environment is essential in corporate organizations and the higher education sector where E-Learning involves knowledge and skills acquisition.

1. Closing the Chasm—
Enhancing Local Knowledge through Spatial Information Technologies

Gernot Brodnig

Regional Centre Bangkok
United Nations Development Programme
gernot.brodnig@undp.org

Viktor Mayer-Schoenberger

Kennedy School of Government
Harvard University
Viktor_Mayer-Schoenberger@harvard.edu

Abstract

Spurred on by the debate on the Digital Divide, recent years have seen a proliferation of initiatives to harness the potential of information and communication technologies (ICT) for the benefit of the poor and disadvantaged. At the same time, development programmes and projects are increasingly trying to supplement, if not substitute, Western science-based approaches with an emphasis on local knowledge. This encounter of modern technology with traditional cognitive systems is particularly prominent in the area of sustainable natural resources management.

This paper looks at a subset of ICT—spatial information systems (including Global Positioning System [GPS] and Remote Sensing), which collect, manipulate and distribute data on a variety of environmental factors, in order to inform and encourage sustainable resources management practices. We examine grassroots and participatory Geographical Information Systems [GIS] approaches, with a particular emphasis on community mapping initiatives in South East Asia. The paper analyses how these projects have empowered local communities by helping them and their counterparts to find a common language and to build a bridge between traditional knowledge and the demands of a globalized world.

1. Introduction

Accurate and reliable information is a key ingredient, if not a precondition, for sustainable development. With the growing importance of the knowledge economy, developing countries need to harness the potential of new technologies and networks.

The revolution in the acquisition, processing and dissemination of information has also had significant repercussions for natural resources management. The challenge to understand the complexities of ecosystems and to mitigate the negative impacts of man-made interventions has spurred a flurry of research programs, networking efforts and scientific co-operation. A veritable glut of data is now at the disposal of policy-makers and planners.

In this quest for more and better information, researchers and decision-makers have recently turned to a long ignored body of knowledge about the environment, that of local people and communities. Particularly those, whose livelihoods depend on the subtleties of ecological cycles and patterns, have accumulated a body of wisdom, commonly referred to as "traditional environmental knowledge" (TEK)[2].

In the past, Western scientists and "experts" tended to regard TEK as methodologically questionable, anecdotal, or—at best—of localized importance. At the same time, the adoption of science-based innovations and technologies by local people has often been stifled by their incompatibility with traditional value systems and cultural practices.

This tension between two epistemologies, which has plagued many development projects, is the central subject of this paper. We examine how new information and communication technologies (ICT) have shaped and altered this relationship. We have chosen to illustrate this question by looking at spatial information technologies (GPS, GIS, Remote Sensing) and their use by local communities in Asia and elsewhere. Our analysis seeks to assess whether these technologies can close the chasm between Western science and local knowledge, and thus contribute to more equitable and sustainable development.

To frame this question, the paper reviews the intrinsic qualities and use of spatial information, both in scientific analysis and traditional environmental knowledge. We then offer a brief synthesis of the main features of spatial information technologies, to be followed by the core of the paper, in which we examine the technologies' advantages and disadvantages for community-based natural resources management.

---

[2]   Other terms in use are "indigenous knowledge" and "traditional ecological knowledge" (Lawas and
       Luning 1996).

2. Two Ways of Knowing

The above-mentioned dichotomy between science and "folk wisdom" applies also to a particular subset of information—spatial or geographic data. These are data that include a reference to a location on the surface of the earth. In the Western scientific tradition, co-ordinate systems such as latitude/longitude are the most common geo-reference. We typically represent spatial data through points, lines or polygons and visualize them in maps. On the other end of the spectrum, local knowledge about the environment is represented in a variety of forms often involving non-linear notions of space, references to stories, myths and different taxonomies. Maps often exist "only" in the minds of the people[3].

2.1 The Science of Space

Accurate and comprehensive spatial data play a critical role in all areas of environmental management and sustainable development. Apart from the various branches of geography, whose disciplinary nerve centre evolves around spatial relations, almost all other "environmental sciences" rely on geographical information. Be it as an input for pollution models of environmental chemists, the gap analysis of population ecologists or the formulas of spatial econometrics, data are acquired and processed at all levels of aggregation.

While in the past most spatial information was the prerogative of governments and government-sponsored research institutions, these days new players—particularly from the private sector—have entered the market. The epitome of old-style spatial analysis, the bound atlas, is quickly being superseded by digital maps and networked databases.

In most developing countries, reliable spatial data are often a scarce resource. This can be attributed not only to the remoteness of many areas but also to the lack of technical capacities and competing priorities in the fledgling economies. Therefore, in many countries, maps and other geographical products from the colonial era often serve as the baseline.

In addition, a number of institutional factors have hampered the use of comprehensive spatial data in environmental planning and management. Fox (1991) notes, for example, that among the most common obstacles to be found in Asia are official restrictions for security reasons as well as the rigidity and compartmentalization of government bureaucracies, which consider certain types of information as their property. Furthermore, in some countries bottlenecks in terms of financial and human resources have hindered the establishment of efficient and effective spatial information systems.

Another feature in the acquisition and use of geographical information has been the dominance of the Western science paradigm, which led to a large-scale dismissal

---

[3] For an example of cognitive/mental mapping see Brody 1981.

of local environmental knowledge and wisdom. As Chambers (1980) observed: "The most difficult thing for an educated expert to accept is that poor farmers may often understand their situations better than he does. Modern scientific knowledge and the indigenous technical knowledge of rural people are grotesquely unequal in leverage. It is difficult for some professions to accept that they have anything to learn from rural people, or to recognize that there is a parallel system of knowledge to their own which is complementary, that is usually valid and in some aspects superior".

2.2 Traditional Environmental Knowledge

It is only in the recent past that decision-makers and members of the scientific community have "discovered" that local people can be environmental "experts", particularly where they have maintained their traditional livelihoods and subsistence economies (Freeman, 1999; Johnson, 1992). Even in the realm of global policy-making, traditional environmental knowledge has received a belated acknowledgment as a legitimate source of information and guidance. Chapter 26 of Agenda 21, for example, calls for the "recognition of [indigenous peoples'] values, traditional knowledge and resource management practices with a view to promoting environmentally sound and sustainable development"[4] (Quarrie, 1992).

What are the specific features of TEK and how does it differ from Western scientific epistemologies? Johnson (1992) defines it as "a body of knowledge built up by a group of people through generations of living in close contact with nature. It includes a system of classification, a set of empirical observations about the environment and a system of self-management that governs resource use". This definition emphasizes a number of structural properties:

- TEK is in essence a geographical information system derived from and embedded in the close relationship of local people with their land and natural resources.
- This knowledge system functions like a distributed database, with members of the community serving as repositories for different types and categories of data, according to their experience and social status.
- TEK is a "scientific" system in that it consists of taxonomic structures, employs particular methodologies and relies on its own "experts".

Like any other knowledge system, TEK is not static. On the contrary, its information base is constantly renewed and revised. Unfortunately, in many cases the dynamics of globalization, industrialization and urbanization have favoured the assimilation and loss of much of this knowledge. This trend has been facilitated by the fact that a

---

[4]   Similarly, Article 8j of the Convention on Biological Diversity calls on the Contracting Parties to "...respect, preserve and maintain knowledge, innovations and practices of indigenous and local communities embodying traditional lifestyles relevant for the conservation and sustainable use of biological diversity and promote their wider application with the approval and involvement of the holders of such knowledge, innovations and practices and encourage the equitable sharing of the benefits arising from the utilization of such knowledge, innovations and practices." (Glowka *et al.* 1994)

significant amount of this information is oral history (Johannes, 1989; Johnson, 1992).

Two Examples of Traditional Environmental Knowledge

- The Akha are a group of mountain peoples spread over southern China, eastern Myanmar, Laos, northern Vietnam and northern Thailand. A body of customary law called zang has traditionally governed them. These rules and regulations include detailed knowledge about the rich local flora and fauna, which informs the management of the village forests. In order to protect their settlements and to have a source of food and medicine, the Akha commonly maintain a forest belt around their villages, even in heavily deforested areas.

- The Marovo people occupy a group of islets and lagoons in one of the seven provinces of the Solomon Islands of the South Pacific. They are fishermen, and their livelihoods depend on the sustainable exploitation of the marine resources. To this end, they have developed approximately 60 different fishing methods. In a documented conversation between a Marovo fisherman and an internationally renowned marine biologist, the local expert was able to teach the Western scientist various subtleties of courtship and spawning among a group of small coral reef fish.

Source: Johnson, 1992.

2.3 The Challenge of Reconciliation

The Brundtland Report, *Our Common Future,* and other key documents of the sustainable development debate have stressed the importance of harmonizing these two "ways of knowing". Agenda 21, for example, notes that governments should "provide local communities and resource users with the information and know-how they need to manage their environment and resources sustainably, applying traditional and indigenous knowledge and approaches when appropriate"[5] (Quarrie, 1992). But how can such integration be achieved? After all, the differences between these two epistemologies seem considerable. Table 1.1 gives a generalized overview of the two systems.

Given these attributes, what role can spatial information technologies play in reconciling TEK and Western science? Like other tools of the information revolution, GIS, GPS and Remote Sensing have seen a rapid proliferation, both in the developed and the developing world. While many observers argue that these technologies contribute to an empowerment of local-level actors and their knowledge systems, others are more sceptical and consider them as part and parcel of Western scientific hegemony. Before taking up these views in more detail later, we need to review briefly the technologies and their key features.

---

[5]   In a similar vein, the Recommendations of the Inuit Circumpolar Conference (1996) on Traditional Indigenous Knowledge and Science call for "collaborative efforts between researchers and communities" and "an atmosphere of co-operation, and not of competition".

Table 1.1
Comparing TEK and Western Science

Traditional Ecological Knowledge	Western Science
• Oral tradition	• Written tradition
• Learned through observation and hands-on experience	• Taught and learned abstracted from the applied context
• Holistic approach	• Reductionist
• Intuitive mode of thinking	• Analytic and abstract reasoning
• Mainly qualitative	• Mainly quantitative
• Data generated by resource users (inclusive)	• Data collected by specialists and experts (exclusive)
• Diachronic data (long time-series on one location)	• Synchronic data (short time-series over a large area)
• Environment as part of social and spiritual relations	• Hierarchical and compartmentalized organization
• Based on cumulative, collective experience	• Based on general laws and theories

Adapted from Johnson, 1992.

3. Spatial Information Technologies—An Overview

Spatial information technologies include three major types:

(i)   Global Positioning System (GPS)

(ii) Geographical Information Systems (GIS)

(iii) Remote Sensing (Airborne and Satellite)

The three categories are related and often employed in an integrated fashion; remote sensing images and GPS data serve as input into GIS, and aerial and satellite images are often verified or "ground-truthed" with GPS co-ordinates. Furthermore, many spatial data are now available in distributed and Web-based databases[6]. Thus, the developments in spatial information technologies cannot be separated from the general trends in ICT.

---

[6]   See for example the GIS Data Depot at www.gisdatadepot.com.

3.1 GPS

The Global Positioning System is a satellite-based navigation system consisting of a network of 24 satellites, which send out continuous signals. On earth, a GPS receiver compares the time a signal was transmitted with the time it was received. This time difference gives the distance of the satellite. By adding distance measurements from two more satellites, the position of the receiver's latitude/longitude can be triangulated (Larijani, 1998; Poole, 1995a).

The accuracy of a GPS position depends on a number of variables. Originally developed for the U.S. military and still managed by it, the GPS system was until 2000 subject to "selective availability", the deliberate degradation of the signal for civilian purposes[7]. All signals are subject to errors due to atmospheric effects, the constellation of the satellites and terrain features such as mountainous or urban areas.

GPS has been used for quite some time in aerial and maritime navigation, and like many other electronic technologies, a gradual miniaturization of components has allowed GPS receivers to become very small and affordable[8]. As a result, uses of the technology have proliferated and include now various forms of civilian and military navigation, mapping and surveying, habitat inventories and wildlife tracking (Larijani, 1998).

Additionally, the European Union has announced its intention to design and deploy a constellation of satellites by 2008, providing an alternative satellite-based navigation system of global reach. The so-called Galileo system will provide accuracy equal to or better than GPS and will be available for unrestricted civilian use.

3.2 GIS

A Geographical Information System is generally defined as a computer-based system used to capture, store, edit, analyse and display geographically referenced data (Poole 1995a). In other words, location information is linked to non-spatial data, commonly referred to as attributes. While "digital maps" can be generated by other software types such as Computer-Aided Design (CAD) programs, GIS is a distinct and more powerful tool. Its main strength is the integration of spatial to attribute data and the layering of information in a number of different themes, which can be superimposed upon each other, revealing complex spatial relationships between variables.

One can distinguish five general functions in a GIS: input, manipulation, management, query and analysis, and visualization (Johnson 1997).

- Input—Spatial and non-spatial data must be translated into a compatible format, which, in the case of geographical data, often involves the digitization of analogue information such as traditional paper maps.

---

[7]   This policy was made obsolete because of techniques such as differential GPS, which provides accuracies of 1m or better.

[8]   Handheld GPS receivers sell for as little as $75.

- Manipulation—Often necessary prior to analysis, in order to make different data compatible.

- Management—Typically based on relational database concepts and internal reference structures.

- Queries and Analysis—The key function and strength of GIS, which includes the creation of new geographical structures through topological overlays and the identification of variables within a given distance of an object or area.

- Visualization—GIS enables the display of maps that can be tailored to specific needs, and be rapidly and dynamically updated.

Geographical Information Systems have become an integral tool in a number of applications, including environmental management and conservation[9]. GIS programs now cover a wide spectrum of complexity and cost, and many universities and private companies offer degrees and training programs in the use of this technology. Big software firms also offer a range of services, and GIS has developed its own industry with trade fairs, specialist journals and a flourishing consultancy circuit.

This is also partly true in developing countries, and tends to be a function of the general penetration of information technologies. While the functionality of GIS should favour widespread adoption, there are, however, a number of obstacles (Yapa, 1991; Yeh, 1991):

- Cost—Despite the drastic decrease in prices of computer hardware and software, for many institutions in the developing world the acquisition and operation of suitable GIS equipment is beyond their reach.

- Complexity—While many GIS packages have become very user-friendly, some of them have exploited the advances in processing power by adding ever-new functionalities. This may require significant investment in training and capacity building, yet another scarce resource for many developing countries.

- Data—A GIS is only as good as the data available. While the GPS has helped to improve the accuracy of many data, the availability of useful digital data is often limited, and, where given, might entail significant cost. While more and more geographical data are available for free on the Internet, they often do not meet the specific needs of users.

3.3 Remote Sensing

A third set of technologies that has had a profound impact on environmental management is Remote Sensing. Aerial photography and videography have been used for a number of applications for a long time. In recent years, two trends have shaped the evolution of these technologies: (i) Digital cameras are rapidly replacing analogue models, facilitating the ease of data manipulation; (ii) Advanced sensors and optics are breaking new grounds in the resolution and quality of images. These advances have made significant contributions in many applications such as tropical forest con-

---

[9]   A good impression of the variety of applications can be gained by looking at the ESRI's (the market leader in GIS software) user showcase at: www.esri.com/user_showcase.html (29 September 2003).

servation (Freeman and Fox, 1994). For example, TREE (Tropical Forest Ecology Experiment), a NASA-sponsored project, successfully tested a series of new sensors. One of them, AVIRIS (Airborne Visible and Infrared Imaging Spectrometer), separates sunlight reflected from the surface into up to 224 wavelengths across the electromagnetic spectrum and can thus depict such properties as vegetation densities, plant chemistry and water contents of leaves (O'Neill, 1993).

Even more significant advances have been brought about by the evolution of satellite technology. Like airborne sensing, the benchmark of development is the image resolution. Landsat 1, the first earth observation satellite, launched by the US in 1972, had a resolution of roughly 80m. In the second half of the 1980s, a major breakthrough occurred with the first French SPOT satellites. SPOT not only brought the resolution down to 10m but also initiated an ambitious program for the commercial distribution of its images. Currently, the highest resolution images come from satellites such as Ikonos and Quickbird[10]. While most remote sensing satellites operate in the visible and infrared electromagnetic spectrum, there are also radar satellites[11], which can penetrate obstacles such as forest and cloud cover. They have been particularly useful in the monitoring of land use, crop conditions and deforestation.

The preceding survey shows that spatial information technologies have taken a firm place in a science-based approach to environmental management and sustainable development. While a number of obstacles to their widespread adoption and use still exist in many developing countries, notably cost and human resources, they have become familiar tools for planning agencies and environment ministries. But what about local communities and their traditional environmental knowledge? Are these technologies widening the gap between the haves and have-nots and reinforcing the prevalence of one development epistemology at the cost of the other? Or are they helping to subvert existing power relations and information monopolies and to empower weaker groups? Table 1.2 lists a number of applications, where rural communities in various land and resource management projects have used spatial technologies.

4. Closing the Chasm?!

It is only in the last decade that researchers and development practitioners have started to explore how spatial information technologies can assist in the integration of traditional environmental knowledge and Western science-based paradigms of environmental management and sustainable development (Aldenderfer and Maschner, 1996; Poole, 1995b; Tabor and Hutchinson, 1994). This debate is often shaped by the growing trend towards bottom-up and participatory models of environmental planning and development (Hutchinson and Toledano, 1993; Kyem, 1998). Further-

---

[10]   See www.spaceimaging.com and www.digitalglobe.com.

[11]   The European Space Agency (ESA) operates ERS-2, and Canada RADARSAT-1.

more, the proliferation of information technologies has resulted in a more reflexive approach *vis-à-vis* their utilization and impacts on society. We can distinguish at least three levels of discourse on the benefits and pitfalls of spatial information technologies. While many of the following arguments and considerations apply equally to the use of GPS and Remote Sensing, our analysis focuses on GIS, as it has attracted the brunt of the pertinent literature and illustrates best the critical issues[12].

Table 1.2
Matching Applications with Technologies

Application	Data Requirements	Technology
Demarcation	Base Maps, Images	GPS
Land use and occupancy	Maps based upon local knowledge and practice	Sketch mapping, GPS
Ecosystem monitoring	Sequential visual data	GPS and aerial video, satellite images
Resource inventories	Local data upon base maps	Aerial video/photo, GPS, GIS
Resource management	Comprehensive ecological and socio-economic data	Aerial video/photo, GPS, satellite images, GIS

Source: Poole, 1995.

4.1 Empowerment vs. Marginalization

This dichotomy reflects the most "abstract" perspective on the relationship between GIS and society. Many representatives in the GIS community consider "their" technologies as inherently democratic and empowering, as they allow an ever-increasing number of users to participate in a process of information collection, processing and dissemination (Harris and Weiner, 1998b; Pickles, 1995). Old prerogatives of the state and governments have been eroded by the proliferation, accessibility and commodification of spatial data. Similarly, prices for hardware, software and data are in a downward spiral, making the technology available to a large number of users.

While this view dominated the early stages of GIS use and dissemination, the past years saw a growing number of more sceptical voices (Fox, 1998; Rundstrom, 1995). Many observers have started to question the positivism and hegemonic power relations embedded in GIS. They emphasize, above all, the differential access to GIS data and equipment. Rather than being a "technology of freedom" (Pool, 1983), GIS and related tools reinforce, they argue, the prevalent technocratic paradigm. As a result,

---

[12]   Moreover, GPS data and aerial/satellite images are increasingly used as input for GIS.

those groups in society that are already disadvantaged are further marginalized, as they cannot participate fully in the benefits of the spatial information revolution.

A third school of thought seeks a middle ground by noting that GIS and related technologies are a double-edged sword and often contradictory. According to them, impacts on social objectives such as equity largely depend on the political, social and economic context, the technologies are embedded in (Harris and Weiner, 1998b). By emphasizing the linkage of technology and the context it is embedded in this third school follows the footsteps of a larger movement of social constructivism of technology (Fischer, 1992; Bijker *et al.*, 1987).

4.2 Participatory and Community GIS

On a more concrete and pragmatic level, the debate about the benefits of GIS has given birth to the concepts of Public Participatory (PPGIS) and Community-Integrated GIS (CIGIS) (Abbot *et al.*, 1998; Harris and Weiner, 1998a; Kyem, 1998; Mbile *et al.*, 2003). Both notions can be linked to the broader discussion on participatory development and community-based resource management. It is most probably fair to say that, at this point in time, PPGIS and CIGIS are efforts to capture these trends in a process rather than distinct methodologies or approaches.

Harris and Weiner (1998b) see the following criteria as constitutive of a community-integrated GIS:

- Likely agency-driven, but not top-down nor privileged toward conventional expert knowledge;
- Assumes that local knowledge is valuable and expert;
- Broadens the access base to digital spatial information technology and data;
- Incorporates socially differentiated multiple realities of landscape;
- Integrates GIS and multimedia;
- Explores the potential for more democratic spatial decision making through greater community participation;
- Assumes that spatial decision-making is conflict ridden and embedded in local politics.

While the project in Kalimantan (see box) fulfilled most of the criteria enunciated by Harris and Weiner (1998b), Sirait *et al.* (2002) noted a number of constraints, which include the political will of the parties involved to recognize different forms of land rights, and the ability of social scientists and mapmakers to accurately reflect the complex relationships of traditional resource management regimes on maps.

Another challenge of participatory GIS approaches is how to scale them up from purely local concerns to reflect regional or national perspectives (Abbot *et al.*, 1998). The advantage of policy-makers being able to deal with local data in a "credible" and "usable" form might be jeopardized by the richness and focus of this information.

Community Mapping in Kalimantan

In many parts of South-East Asia, forestry is shaped by the tension between state-controlled forest regimes and traditional management practices by local people. In recent years, there has been a growing acceptance of indigenous knowledge and practices, and countries like the Philippines or Indonesia, have recognized customary tenure regimes.

While this represents an important step towards more equitable resource management, in most cases this recognition remained a hollow shell, as land and resource rights were not demarcated. Governments often lack the political will and resources to carry out these processes. In a few projects, however, local communities, with the support of NGOs and researchers, have started autonomous mapping exercises, making use of a variety of spatial information technologies.

One example of such a community-mapping project is the demarcation of customary lands in the village of Long Uli in East Kalimantan. The village territory borders a nature reserve, and in the past, a number of land use conflicts between the villagers and the forest department occurred.

The project's objectives were to:

- Map the customary lands using oral history, traditional knowledge, sketch maps and a global positioning system;
- Use a geographic information system (GIS) to overlay this information with official land use maps in order to clarify land-boundary conflicts; and
- Identify management alternatives taking into account TEK.

The project demonstrated that spatial information technologies can assist in the assessment of indigenous ways of organizing and allocating space, in the reconciliation of local resource management systems and government-instituted science-based regimes, and in the formal recognition and protection of customary forest tenure arrangements.

Source: Sirait et al., 2002

4.3 Culturally-Appropriate GIS

The cultural appropriateness or sensitivity of GIS informs a third perspective on the use of spatial information technologies by local communities. Many GIS practitioners, including indigenous experts, argue that GIS and related technologies can help demonstrate the close relationship between local people and their land by illustrating the multiple dimensions of human-land relations such as folk taxonomies of flora and fauna, place names, myths and legends (Johnson, 1997; Poole, 1995b). The overlay feature and linkages to attribute databases, including multimedia (recordings, photographs, animations, etc.), represent a step away from the classic cartographic metaphors and spatial precepts established by Western science. The easy manipulation of data also makes it possible to integrate and revise input from a large and diverse group of informants.

On a pragmatic level, advocates of spatial information technologies argue that these tools are well suited to preserve, revitalize and disseminate knowledge and practices that are on the brink of extinction (Harmsworth, 1999). Even if a GIS cannot capture all facets of this knowledge, a combination with other methodologies such as participant observation allows a fair representation of local cultures. Furthermore, in order to deal effectively with government authorities and other stakeholders, a common language must be found (Sirait *et al.*, 2002)[13].

Conservation and Dissemination of Maori Traditional Knowledge

Like most indigenous peoples, the Maori of New Zealand have a comprehensive body of traditional knowledge and values, most of which is closely associated with their lands and natural resources. According to present legal requirements, this knowledge must be taken into account in land-use planning. In recent years, there has been a growing interest on the part of the Maori in recording their traditional knowledge, particularly at the local and community levels. With the increased access to computers, these efforts used database management systems to store and organize the information.

Many of these databases have been integrated into geographical information systems and organized according to specific tribal areas. The main challenge of these georeferenced TEK systems was to combine Maori cultural values with biophysical information for the benefit of environmental management planning, while at the same time protecting the confidentiality and intellectual property of certain types of information. The latter goal was accomplished by linking information that is too sensitive to store in the GIS via a database directory to an individual person.

These projects in New Zealand address the critical question as to what degree GIS and related technologies can be utilized in a culturally sensitive and appropriate way and actually help to conserve and disseminate the rapidly vanishing traditional knowledge of local people.

Source: Laituri, 1998 and Harmsworth, 1999.

This pragmatism stands at the core of many criticisms of GIS for local communities. Certain purists consider GIS technology as "a tool for epistemological assimilation, and as such, as the newest link in a long chain of attempts by Western societies to subsume or destroy indigenous cultures" (Rundstrom, 1995). Spatial information technologies cannot capture the "ethnological content of spatial patterns" (Fox, 1998), i.e., the cultural patterns imbued in landscapes and natural resources.

Most traditional resource management practices are characterized by a high degree of fluidity and flexibility. Boundaries that exist in cognitive/mental maps are continuously redefined. Once a map establishes clear borders, claims and entitlements are "out in the open", often rendering the management of conflicts more difficult (Fox, 1998; Sirait *et al.*, 2002). Furthermore, spatial information is often restricted or secret. The knowledge of sacred sites or certain practices is predominantly limited to a small group of people and cannot be disclosed. While the example of the Maori shows that mechanisms can be developed to take privacy considerations into account, spatial information technologies could contribute to violations of this "space".

---

[13] Native groups in Canada and the United States have most persuasively made this argument in a number of land rights claims.

5. Conclusions

Are spatial information systems appropriate technologies (Yapa, 1991)? In other words, can they help local communities to manage their environment more effectively and efficiently, while at the same time respect and enhance traditional cultures? We are tempted to conclude with Krantzberg's First Law: "Technology is neither good nor bad, nor is it neutral" (Castells, 1996), which points to the importance of the social, political and economic context of technologies.

While spatial technologies are an important step towards integrating data at various levels of analysis and from different epistemologies, they can not relieve the researcher and policy-maker from the task of determining which social, economic, and political factors impact human-environment relations and lead to sustainable development. As Abbot *et al.* (1998) observed: "GIS ... is only as good as the local politics".

We hope, however, that the preceding analysis has highlighted another point— The chasm to close between Western science and local knowledge might not be very big. Moreover, given TEK's distinct qualities—embedded in the close relationship of local people with their land and natural resources, structured like a distributed database, and creating a system consisting of taxonomic structures, unique methodologies and experts—traditional environmental knowledge seems in many ways much closer to the technologies' structural features and functionalities than some of our Western methodologies.

Whatever the exact repercussions of GIS, GPS and Remote Sensing might be in the individual case, one thing seems to be sure—Even in the age of the Internet, cyberspace and virtual reality, there is room for space.

References

Abbot, J., Chambers, R., Dunn, C., Harris, T., Merode, E. de, Porter, G., Townsend, J. and Weiner, D. (1998) Participatory GIS: Opportunity or Oxymoron?, PLA Notes, 33, 27–34.

Aldenderfer, M. and Maschner, H. (eds.) (1996) *Anthropology, Space, and Geographic Information Systems.* New York: Oxford University Press.

Bijker, W., Hughes, T. and Pinch, T. (1987) *The Social Construction of Technological Systems.* Cambridge: MIT Press.

Brody, H. (1981) *Maps and Dreams.* Prospect Heights: Waveland Press.

Castells, M. (1996) *The Rise of the Network Society.* Oxford: Blackwell.

Chambers, R. (1980) The Small Farmer is a Professional, *Ceres*, **13**, 2, 19–23.

Fischer, C. (1992) *America Calling. A Social History of the Telephone to 1940.* Berkeley: University of California Press.

Fox, J. (1991) Spatial Information for Resource Management in Asia: A Review of Institutional Issues, *International Journal for Geographical Information Systems,* **5**, 1, 59–72.

Fox, J. (1998) Mapping the Commons: The Social Context of Spatial Information Technologies, *The Common Property Resource Digest,* **45**, 1–4.

Freeman, M. (1999) The Nature and Utility of Traditional Ecological Knowledge. www.carc.org/pubs/v20no1/utility.htm.

Freeman, P. and Fox, R. (1994) *Satellite Mapping of Tropical Forest Cover and Deforestation: A Review with Recommendations for USAID.* Arlington: ENRIC.

Glowka, L., Burhenne-Guilmin, F. B. and Synge, A. (1994) *A Guide to the Convention on Biological Diversity.* Gland: IUCN.

Harmsworth, G. (1999) Indigenous Values and GIS: A Method and a Framework, *Indigenous Knowledge and Development Monitor,* **6**, 3, www.nuffic.nl/ciran/ikdm/6-3/harmsw.html.

Harris, T. and Weiner, D. (1998a) Community-Integrated GIS for Land Reform in Mpumalanga Province, South Africa. www.ncgia.ucsb.edu/varenius/ppgis/papers/harris.html.

Harris, T. and Weiner, D. (1998b) Empowerment, Marginalization and 'Community-integrated' GIS, *Cartography and Geographic Information Systems,* **25**, 2, 67–76.

Hutchinson, C. and Toledano, J. (1993) Guidelines for Demonstrating Geographical Information Systems Based on Participatory Development, *International Journal for Geographical Information Systems,* **7**, 5, 453–461.

Inuit Circumpolar Conference (1996) Recommendations on the Integration of Two Ways of Knowing: Traditional Indigenous Knowledge and Scientific Knowledge.

Johannes, R. (ed.) (1989) *Traditional Ecological Knowledge: A Collection of Essays.* Gland: IUCN.

Johnson, B. (1997) The Use of Geographic Information Systems (GIS) by First Nations. www.nativemaps.org/abstracts/ben.html.

Johnson, M. (ed.) (1992) *Lore: Capturing Traditional Environmental Knowledge.* Ottawa: Dene Cultural Institute.

Kyem, P. (1998) Promoting Local Community Participation in Forest Management through the Application of a Geographic Information System: A PPGIS Experience from Southern Ghana. www.ncgia.ucsb.edu/varenius/ppgis/papers/kyem/kyem.html.

Laituri, M. (1998) Marginal Societies and Geographic Information Systems. Empowerment, Marginalization, and Public Participation GIS Meeting Santa Barbara, California 14–17 October www.ncgia.ucsb.edu/varenius/ppgis/papers/laituri.html.

Larijani, L. (1998) *GPS for Everyone.* New York: American Interface Corp.

Lawas, C. and Luning, H. (1996) Farmers' Knowledge and GIS, *Indigenous Knowledge and Development Monitor,* **4**, 1. www.nuffic.nl/ciran/ikdm/4-1/articles/lawas.html.

Mbile, P., De Grande, A. and Okon, N. (2003) Integrating Participatory Resource Mapping and Geographic Information Systems in Forest Conservation and Natural Resources Management in Cameroon: A Methodological Guide, *The Electronic Journal on Information Systems in Developing Countries,* **14**, 2, 1–11,

O'Neill, T. (1993) New Sensors Eye the Rain Forest, *National Geographic,* **189**, 6, 118–130.

Pickles, J. (ed.) (1995) *Ground Truth: The Social Implications of Geographic Information Systems.* New York: Guilford Press.

Pool, I. (1983) *Technologies of Freedom: On Free Speech in the Electronic Age.* Cambridge: Belknap Press of Harvard University Press.

Poole, P. (1995a) Guide to the Technology, *Cultural Survival Quarterly,* **18**, 4, 16–18.

Poole, P. (1995b) Indigenous Peoples, Mapping and Biodiversity Conservation: An Analysis of Current Activities and Opportunities for Applying Geomatics Technologies. Peoples and Forest Program Discussion Paper. Biodiversity Support Program, Washington, D.C.

Quarrie, J. (ed.) (1992) *Earth Summit 1992.* London: Regency Press.

Rundstrom, R. (1995) GIS, Indigenous Peoples, and Epistemological Diversity, *Cartography and Geographic Information Systems,* **22**, 1, 45–57.

Sirait, M., Prasodjo, S., Podger, N., Flavelle, A. and Fox, J. (2002) Mapping Customary Land in East Kalimantan, Indonesia: A Tool for Forest Management.
www2.eastwestcenter.org/environment/fox/kali.html.

Tabor, J. and Hutchinson, C. (1994) Using Indigenous Knowledge, Remote Sensing and GIS for Sustainable Development, *Indigenous Knowledge and Development Monitor,* **2**, 1.
www.nuffic.nl/ciran/ikdm/2-1/articles/tabor.html.

World Commission on Environment and Development (1987) *Our Common Future.* Oxford: Oxford University Press.

Yapa, L. (1991) Is GIS Appropriate Technology?, *International Journal for Geographical Information Systems,* **5**, 1, 41–58.

Yeh, A. (1991) The Development and Application of Geographic Information Systems for Urban and Regional Planning in the Developing Countries, *International Journal for Geographical Information Systems,* **5**, 1, 5–27.

About the Authors

Dr. Gernot Brodnig is the Environment Policy Advisor for Asia, based at UNDP's Regional Centre Bangkok. Prior to that, he was a Research Fellow at Harvard University's Kennedy School of Government and working on the policy implications of spatial information technologies.

Dr. Viktor Mayer-Schoenberger is Associate Professor of Public Policy at Harvard's John F. Kennedy School of Government, where he specializes on public policy issues of information and communication technologies. He is also on the editorial board of "Information Technology for Development".

2. Endogenizing IT Innovation—
Preliminary Considerations for an Approach to Socio-Economic Development

Stephen S. Corea

Department of Operations Research and Systems
Warwick Business School
University of Warwick
Coventry CV4 7AL

steve.corea@wbs.ac.uk

Abstract

This theoretical paper undertakes to conceptualize the notion of technological innovation and societal development in compatible terms, in order to elaborate an innovation-based approach to promoting the socio-economic growth of societies—one with a central role for the use of information technologies. A behavioural notion of development, based upon the structural conditioning of individual and group behaviours, is found congruent, and useful, for understanding the dynamics of transformation by which a fundamental orientation towards practices of technological innovation can significantly raise the level of societal wealth over the long term. Insights are recruited from sociotechnical systems theory to inform a preliminary elaboration of principles by which social systems may be organized to generate learning behaviours that lead to continuous improvements and socio-economic prosperity. The main outcome of this paper's exploration is to develop the argument that technological innovation in less advanced societies should constitute a self-cultivated rather than an imported behavioural phenomenon. It is of greater priority for such societies to invest in cultivating the patterns and practices of behaviour that underpin the various IT based innovations of modernization, than to pursue alternative policies such as aggressive IT adoption or rapid technical change.

1. Introduction

Many nations across the globe have committed information and communication technologies for the betterment of their societies. Distinctly diverging densities of uptake and incursion have emerged however, generally determined by the affluence, capacities, value systems, and market dynamics of specific nations. These inequities in comparative adoption rates beg particular concern for developing or less developed countries with low rates of IT ingression. These countries may be seen to be missing out on the key potential of computer technologies to promote product and service quality (Davenport, 1993), and enable profitable membership in a global economy increasingly perforated by technology based information flows (Odedra-Straub, 1996). In stark contrast to a small number of industrialized and newly industrialized nations that have been steadily capitalizing on the swift pace of IT innovation, a large number of low-income countries have been prevented by scarcity, socio-political instabilities, or depression of their economies, from actively participating in the benefits of this technological revolution.

The need for policy makers and managers in developing countries to focus on promoting large-scale IT adoption, so that societal benefits will permeate to equivalent degrees of saturation as exist in industrialized countries, has been advanced in the current discourse on societal development (e.g., Talero and Gaudette, 1995). Such counsel often rests on the implicit assumption of the significant or decisive role of computerization in promoting economic growth. Contemporary research on the relationship of IT investment and economic productivity, popularly known as the "productivity paradox", has drawn mixed and conflicting results. For instance, a major study by Dewan and Kraemer (2000), based on data from 36 countries, concluded that growth in IT investment is positively correlated with productivity growth in general. Avgerou (1998) however, in a review of theoretical literature addressing the relationship between IT and economic growth, found no substantive support for the premise that investment in ICT technologies leads directly to economic prosperity. Similarly, Quah (2000) found a negative relationship between IT investment and US productivity growth in the period of 1964 to 1993. Productivity in the US was high when IT investment was low during the 1960's, and dropped steeply when IT investment swelled after 1972.

However, while considerable research effort has been focused on identifying the exact relationship between IT and economic productivity, less attention has been given to conceptual elaboration of the broad relationship between IT based innovation and societal development. Hence, this theoretical paper undertakes sets out to explore this relatively untreaded track, one that might provide alternative ways of evaluating the proposition of achieving growth through IT adoption. An attempt is made to conceptualize the notions of technological innovation and societal development in congruent terms, so as to support a preliminary elaboration of an IT innovation-based approach to promoting the socio-economic development of societies. It is hoped the implications drawn in this exploration will offer policy-makers and managers some ideas or insights that may prove beneficial for understanding the engineering of sustained developmental growth, and the reduction of technological disparities.

2. Conceptualizing Development through Innovation

This section undertakes an elaboration of a technological innovation-based notion of societal development. It first introduces key terms and notions that form the conceptual scaffolding for subsequent discussions (section 2.1). This is succeeded by a preliminary conceptualization of the association between technological innovation and development (section 2.2).

2.1 Conceptual Foundations

This section introduces terminology and ideas from several theoretical bases that will facilitate subsequent exploration of innovation and development. Relevant notions are covered from development theory, structuration theory, cybernetics, the technological and IT innovation literature.

2.1.1 A Behavioural Definition of Development

The term development requires elucidation. A distinction may be made between the goal and process of development. Contemporary theoretical notions of the objective of development emphasize a multi-composite transformation in the phenomenological embedding of individuals within a society, involving a considerable shift away from a condition of life summarily perceived as unsatisfactory towards one that is significantly preferable in both spiritual and material dimensions (Owens, 1987; Todaro, 1994). This holistic viewpoint will be adopted for this paper's purpose. Adapting from Todaro's (1994) definition, the objective of development will be held to encompass the acceleration of economic growth, the reduction of inequality, and the eradication of poverty. Much work has been produced on the process of development (Todaro, 1994). This literature has been dominated by five major economic models: the linear stage-growth theory (1950s–60s), structural change theory or international dependence model (1970s), the neo-classical counter-revolution (1980s), and new growth theory (1990s).

Rather, for the purpose of the analysis to follow, a more generic and sociological formulation of the process of growth is needed, one which excludes the bias of political ideologies that inform the above models (see Rapley, 1996), and which also encompasses various non-economic dimensions of growth, including social, cultural and psychological factors. Such sociological formulations of societal development have tended to fall into two categories—psychodynamic models and behavioural models. Psychodynamic models emphasize the psychological make-up of individuals in society (e.g., values, attitudes, personality) as the principle driving force behind the aggregate socio-economic behaviour that characterizes a nation's development[14]. However, the applicability of psychodynamic formulations has been keenly undermined by serious problems with the empirical validity of such studies, as well as the existence of significant data and findings that demonstrate rapid changes in societal behaviour difficult to explain on the basis of such models (Kunkel, 1970). Moreover, the premise underlying these models, that value change has to precede behaviour change, is an idealistic assumption that fails to consider the play of situational context and exogenous variables in behaviour alteration (Hofstede, 1984).

Kunkel (1970) proposed a behavioural model that reflects the second category of sociological explanations of human behaviour underlying societal phenomena. Behavioural models, traceable to the work of earlier sociologists like Homans (1961) and Blau (1964), are strictly characterized by their underlying basis in some variant of learning theory. The fundamental premise of such models is that behaviour is learnt by social conditioning, and *can be modified* at any time by changing the circumjacent operant conditions (Homans, 1961). Correspondingly, Kunkel asserts that the key challenge of economic development is the identification and change of those selected aspects of man's social environment that are relevant to the learning of new behaviour (1970, p. 76). This formulation of the process of societal development as the alteration of structural conditions conducive to learning is particularly apt for this paper's aim of establishing a correspondence with technological innovation or change[15]. Technological change is held by contemporary economists to be the most important driver of economic growth and social prosperity (Thomson, 1993b), and is fundamentally characterized as a *learning* process (Thomson, 1993a). To facilitate upcoming elaborations of a technological innovation-led notion of societal development, the following definition of development, derived from Kunkel (1970), is adopted—a multidimensional process, involving significant changes in selected aspects of the social environment of individuals or groups that are necessary for the learning of new behavioural repertoires to accomplish the acceleration of socio-economic growth, the reduction of inequality, and the eradication of poverty.

---

[14]   Important works in this stream include: (i) a classic study by McClelland (1961), which emphasized the need-achievement factor of individuals in different cultures as being responsible for national differences in economic prosperity; and (ii) one by Ayal (1963), that traced the differences in economic progress between Japan and Thailand to differences in their modal cultural value systems.

[15]   The terms technological innovation and change are used interchangeably in this paper, although some might view innovation in a more narrow sense as being a subset of technological change in general. Moreover, references to technological innovation or change in this paper imply information and communication technologies in general, although the notion of technology adopted in the innovation literature cited in this paper typically encompasses a much broader set than solely computer-related technologies.

2.1.2 Structural Conditioning and Change

This notion of development advanced above advocates changes in social structure as engendering changes in individual or group action. This relationship between structure and behaviour is a major domain of inquiry in the social sciences. A major issue of debate has been an apparent contradiction between those theories that focus on the autonomy of human agency or behaviour, and those that focus on the conditioning, often constraining effects of social structure on human action. A sophisticated attempt to resolve this contradiction has been structuration theory (Giddens, 1979; 1984), which posits the relationship between social structure and human action in the form of a dualism, in which both phenomena are recursively-defined. Social structure is perceived as both shaping (i.e., enabling and constraining) human actions though the provision of modalities, and as being shaped and reproduced at the same time by the unfolding of human action (Giddens, 1979; 1984). Three modalities are identified by Giddens (1984), as mediating the influence of structure on behaviour as well as vice-versa: (i) interpretive schemes, which are socially-shared knowledge bases that individuals use to communicate and interpret events and actions; (ii) resources, which are the means by which actions are accomplished and goals realized; and (iii) norms, which refer to social rules and values that channel human behaviour. Structuration theory is useful as a meta-framework for studying IT innovation practices in social systems (Orlikowski, 1996).

The notion of development adopted in the previous section may be informed and qualified by structuration theory. Thus, in advocating the change of selected aspects of social structure as a way to promote changes towards development-oriented behaviour, due recognition is forthcoming that this will also invoke a reshaping of those aspects of structure by the unfolding of action. Precedence of focus, however, will be given here to the conditioning influence of structure over action. The notion of development endorsed here thus advocates selected changes in social structure that reconfigure the availability and accessibility to individuals (or groups) of key interpretive schemes, resources and norms that promote the learning of new behaviours conducive to societal development.

The notion of development advanced above is constituted by two types of change: (i) changes in structural aspects of the social environment, that are relevant to (ii) changes in behaviour (i.e., learning new behaviours). Since change is an important concept, a greater understanding of its dimensions is necessary to develop the upcoming elaboration.

Two main forms of change can be distinguished in social systems—morphogenesis and morphostasis (Smith, 1982). Morphogenesis refers to a form of change which, either through small incremental stages or larger radical transformations, alters the structure of an entire social system so completely that it is never the same again. The transformed system then persists for a time in its new state, until a new total transfiguration occurs. This repeats over successive cycles. Morphogenic change is thus change that triggers more changes—it is autocatalytic. Systems that may be described as morphogenic in character thus reflect a fundamental tendency towards disequilibrium. Morphostasis, on the other hand, refers to change that is superficial. It is typically short-lived, disappearing unless continual effort is devoted to maintaining its semblance. It may even manifest as a kind of paradoxical guise, appearing to enable an appreciable difference, while on deeper consideration, things are found to remain essentially the same as ever. Thus, morphostatic change implicates the power of inertia in systems—a preservation-oriented tendency in which countervailing forces absorb or overcome initial perturbations to return a system to its original equilibrium (or homeostasis). These concepts will be used analogously later.

2.1.3 Technological Innovation

Technological change, considered in this paper as synonymous with innovation, has been defined as any incremental or radical changes in the application of problem-solving knowledge to the production process, resulting in increased efficiency, either in the form of a product or service produced with lower costs, or in the form of a qualitative improvement in the nature of a product or service (Mokyr, 1990; Chiaromonte and Dosi, 1993).

The critical role of technological change in driving the long-term improvement of the national economic performance of countries is widely acknowledged (Rosenberg, 1994), and similar importance has been accorded to it at the meso-level of the growth of business corporations (Thompson, 1993a). Only in recent decades, however, have social scientists started to study the processes of technological innovation, rather than treating it merely as a black-box of growth input (Rosenberg, 1982) i.e., as a given. In doing so, they have encountered a very rich phenomenon that is too complex to fit easily into heuristic models (Thomson, 1993a). The characteristics of this phenomenon apply indiscriminately to all kinds of technologies, including information technologies, which are critical tools of modern process innovation (Davenport, 1993). Some of the important conclusions that have dominated recent studies of the characteristics of technological innovation include the following: (i) that it is an ever-ramifying network of minor and major innovations, involving the application of public and private knowledge bases (Rosenberg, 1994); (ii) that it entails a process of cultural evolution (Nelson, 1993); (iii) that its dynamics of improvement are evolutionary and iterative, rather than linear (Smith, 1993); (iv) that it tends to follow certain trajectories, defined by specific paradigms of knowledge and expertise loosely linked to scientific principles and processes (Dosi, 1988); (v) that the progressive exhaustion of its trajectory triggers embarkation in new ones (Chiaromonte and Dosi, 1993); (vi) and that it is path-dependant, in that its current and future outcomes are dependent on the effects of past events (David, 1993).

The thinker Jacques Ellul (1964) elaborated the ideological basis that may be seen to underlie technological innovation. He portrayed technology in this modern age as a powerful, autonomous and self-perpetuating force. He termed this force "technique", and suffused it with several meanings: (i) the totality of practices that impels towards absolute efficiency as its ultimate goal; (ii) a universally conquering, all-transforming, monolithic phenomenon, which has become an end-in-itself; (iii) a form of rational intentionality actualized in machines. Its ethical implications aside, Ellul's work highlights some of the salient features of technological change.

Above all, technological change is constituted by a learning process, involving, in varying degrees, stages/steps of discovery, experimentation, development, imitation and adoption (Thomson, 1993a; Dossi, 1988). Technological innovation is thus founded on behaviour that involves a learning and problem-solving orientation. Mokyr (1990) defined the characteristics of *innovative behaviour* in individuals and social groups (i.e., behaviour resulting in technological change) as a willingness to challenge problems perceived in the environment, using ingenious means and whatever resources available, for the sake of improvement and anticipated rewards. Such rewards, realized through the construction of artefacts that successfully address problems, can then serve as incentive for further acts of innovation. Technological change may thus be regarded essentially as learning behaviour that impels toward ever increasing economic output and social value. It may be seen to possess a self-perpetuating momentum, amplifying progressively through endless chains and altering completely the social system that cultivates it. The form of transformation it promotes corresponds completely to that of morphogenesis. As a universally conquering phenomenon (Ellul, 1964), technological innovation appears to be inexhaustible. Studies on the microeconomics of innovation have suggested that unexploited opportunities exist permanently, and that the known is but a small subset of the unknown (Rosenberg, 1982; Dossi, 1988). Technological innovation in a social system is thus potentially capable of triggering unlimited cycles of morphogenic change in the system.

2.2 Technological Innovation and Societal Development

The preceding elaborations regarding the nature of societal development, change and technological innovation provide a context for undertaking a conceptualization of the relationship between technological innovation and societal development. The notions of development and innovation that have been elaborated may be perceived to be essentially consonant. The behavioural basis of the notion of technological innovation is consistent with the behavioural basis of the notion of development in that, for both, learning behaviour aimed at improvement is the key mechanism.

The process of technological innovation may thus be folded in within the compass of development. In a social system structurally geared towards technological innovation, a distinct kind of behaviour is significantly encouraged and generated—behaviour that aims at improvement. In a society with appropriate modalities of structure (i.e., interpretive schemes, rules and resources), such behaviour can be autocatalytic. Behaviour aimed at improvement can result in changes in knowledge/understanding that materialize in new processes or technological artefacts, which then serve as the basis and cause for further behaviour that is aimed at improvement. Technological change can thus be seen as the *principle* reason why industrialized nations, that have learned to harness self-perpetuating systems of innovation, have in the course of history moved further towards achieving the goals set out in the notion of development stated earlier—the acceleration of economic growth, the reduction of inequality, and the eradication of poverty (Mokyr, 1990).

It thus follows that in social systems which are not geared towards technological innovation, stagnation occurs, because behaviour that produces improvement is not being generated. In other words, it may be argued that a low state and poor rate of improvement in competence at technological innovation, both at the level of knowledge or technique as well as artefact construction, is what keeps societies undeveloped, not the lack of capital or infrastructure. For this reason, Kunkel (1970) concludes that low income nations and poverty-stricken communities may be characterized as being morphostatic systems, while industrial nations are typically morphogenic. Such morphostatic social systems are not a reflection of the failing of the people that compose them however. Rather, it may be said to be a failing of the structure or architecture of the social system itself in not producing behaviour among its members that breaks its inertia. Such societal units however, some of whom are locked into the crippling stasis that characterizes the "culture of poverty" (Goulet, 1971), are far from being foregone. Relevant changes in critical aspects of their social environment can produce new sets of behavioural repertoires aimed at improving the socio-economic state of their system. In some cases, transformations may be achieved relatively quickly, as some newly industrialized countries have demonstrated. In others, chronic entropy in their societies may take much longer to undo, but positive change is still feasible. Modifications can swing the architecture of a social system towards a strong orientation to technological innovation.

Social scientists have stressed the pivotal importance in poorer or less developed societies of cultivating a morphogenic impetus in their systems of national development, through technological innovation (Lazonick, 1993; Porter, 1990). Less advanced nations would benefit from making technological innovation a modal or primary behaviour in their societies. Adopting a behavioural formulation of development, individuals and groups in a society could be attuned towards participating more productively in the generation of technological innovation, if changes were made in critical aspects of their social milieu that generated the learning of new behaviours for propagating continuous improvements in the conditions of living.

Since the notion of technological innovation, which represents a technical process (involving problem-solving methods) on one level, has in the course of this section come to be folded in with the notion of development, which on another level represents a social process, some means might be sought by which to harmonize the two opposing "tensions" of social and technical, when introducing changes to gear a society towards technological innovation. A school of thought which offers insights on this is sociotechnical systems theory, reviewed next. The need has been stressed for adopting a sociotechnical approach when designing national systems of development. Boyer (1988) found that major periods of socio-economic growth in society, as well as phases of prolonged dysfunction or sudden downturn, can usually be traced to the compatibility or the mismatch over time between the technical aspects of various subsystems and collective social factors. He warned against analyzing social and technical elements separately since their dynamic interrelationships called for an integral approach. Chiaromonte *et al.* (1993) similarly indicated the need for a sociotechnical focus, when they concluded that economic growth requires a strong coupling between technological learning and behavioural adjustments in key institutional forms of society.

3. Promoting IT Innovation in Societies

Having undertaken a conceptualization of societal development and its dynamics of transformation through the promotion of technological innovation as a modal behaviour, this section seeks to obtain further insights from sociotechnical systems theory that may be used to inform a preliminary elaboration of an approach for harnessing IT innovation as a spur to societal development. The sociotechnical approach is described next (section 3.1), followed by a consideration of implications that may be derived from the preceding conceptual exploration (section 3.2).

3.1 Sociotechnical Systems Approach

Sociotechnical systems (STS) theory offers a visionary blueprint for conceiving how the human and technological components of organized activity systems may best be framed, designed and managed in a synergistic way (Pasmore, 1995; Eijnatten, 1993). In its classical formulation STS theory treats the organizational or societal unit as a complex whole, composed of two independent sub-systems that interact in reciprocal influence (Emery and Trist, 1960)—a social system, comprised of people with social, psychological and biological needs, and a technical system, comprised of tasks, methods, technological artefacts, and the knowledge they embed (Pasmore, 1988). This complex whole must be vigilantly managed in "open" commerce with influential forces and circumstances in its external milieu.

The underlying perspective is that the viability and effectiveness of an organized workforce hinges on how well its technical and social subsystems have been designed or managed in regard to each other, and collectively tuned with respect to the demands of its environment (Emery and Trist, 1960). The most effective way to organize and manage social systems is to bring into existence production or service arrangements in which the requirements of technical components and the human needs of individuals or groups are reconciled in joint optimization. Motivational issues are a primary concern in the design of structures and processes. Emphasis is given to the need to mobilize human motivation and commitment in the process of implementing environments of technology adoption (Mumford, 1996). This need is addressed by making stakeholder participation processes a central input in such endeavours, and work enrichment an underlying design goal. The emphasis on instituting structurally correlated changes, that enrol the motivational drives of individuals and groups in adapting to or exploiting the use of technology, is congruent with the earlier stated idea of societal development as changes in aspects of the social system that foster adaptation to (i.e., learning) new behaviours.

Recently, a critical reformulation of the traditional design philosophy of the STS approach was introduced by its Dutch school of thought (Sitter *et al.*, 1997; Eijnatten and Zwaan, 1998). This change is paradigmatic. It obtains from a stance that the conventional analysis of an organized activity system under separate "social" and "technical" categories inherently contradicts the essence of such a system as an *integral* functional unit (Sitter *et al.*, 1997). Such division is seen to thwart, rather than foster, a coherent understanding of structural dynamics. It can only lead, at best, to a nominal mapping between human and technological components, in which little of real value can be said about the relationship between the ideas, beliefs or norms of the members of a workforce, and the artefacts that they use, such as PCs, software or network infrastructure (Sitter *et al.*, 1997). The idea of designing an activity system by discriminating exclusively social elements from discrete technical ones (and matching them) is rejected, as functional phenomena are concurrently social and technical in aspect (Sitter *et al.*, 1997, p. 504): "A (sub)system is always a complete set of relations between various elements that perform a certain function. Such relations are always social as well as technical." An alternative design philosophy, that emphasizes integral coherence and internal consistency, is offered. The relationship of social and technical aspects can only be appreciated and optimized within a convening of the complete system (Eijnatten and Zwaan, 1998). This emphasis on indivisibility of social and technical is paramount. It sets this approach apart from other philosophies of activity design such as business process reengineering, that often fail to preserve vital coherencies when applied to the development of social systems (Matthews, 1997).

It follows, correspondingly, that the central problematic of the sociotechnical design approach is the integral architecturing or configuring of a system's structure, and how this modulates the opportunities for co-ordination, adaptation and innovation (Sitter *et al.*, 1997). The design and management of a social system should be directed towards fostering and amplifying such opportunities, by modifying the structures of the system to be consistent with their generation.

As a consequence of this reformulation, technology is no longer seen as a separate sub-system by itself. Instead, in a radical shift, technology is conceived as being part and parcel of the "attribute" structure of humans (i.e., individual or group members of a social system), in the same ontological domain as ideas, beliefs, values and norms (Eijnatten and Zwaan, 1998). This explicit recognition of technology as a human subset, belonging to the very same expressive substrate of human nature as ideas, attitudes, values and norms, highlights its fundamental essence as an ideological construct (Kuhns, 1971). It points to the role of expressive human investment (knowledge and aspirations) as the genesis of their innovation, and potent qualifiers of their use. It encourages the consideration of the use of technology in undivided relation to the operation of structurational modalities of human agency, such as interpretive schemes, rules and resources. By annexing IT to the locus of workplace expression, this reformulation in STS design theory points to the need to consider the context of human activity as the rightful qualifier of the appropriateness of a particular IT system or tool (Avgerou and Land, 1992).

3.2 Endogenizing IT Innovation

Given these insights and the prior conceptualizations, certain implications for the design of societies which are principally geared towards technological innovation may be derived. Essentially, it appears important, in the light of the preceding deliberations, to assert the need for less developed or developing societies to develop the competency of their citizens to engage in IT based innovation practices as an *endogenous* behaviour. This is elaborated next.

Technological innovation has thus been seen to involve learning behaviours, which when generated in sufficient mass, are capable of morphogenic transformations of a society or social system towards higher states of material and process improvement. In the long-run, the generation of such behaviours may be argued to demand a greater priority of investment than alternative approaches, such as the aggressive uptake of information technologies imported from abroad, or a strategy of implementing fast technical change. It is more critical to cultivate the set of behaviours that sustains continuous improvement, and that materializes in the technological artefacts and qualitative improvements of life in industrialized nations. This may be seen as more important an issue to address than the large-scale adoption of such costly technologies by societal units whose lack of structural orientation towards change or improvement (i.e., inertia) can render such appropriations morphostatic at most[16]. For the adoption of IT systems among organized activity systems can create only superficial or transient states of flux and performance gain, in which new IT based processes come to be seen over time as being out of synchronization with the older institutionalized body of arrangements, causing organizations to return to ingrained, ineffective ways of functioning, once such new tools and associated work processes have been absorbed into the previous web of calcified performance inefficiencies (e.g., Madon, 1993). Rather than thus foisting such IT systems or tools unsystematically on members of less advanced or poorer societies, the concerted nurturing of behaviours innately adapted to the transformative potential of such tools may prove a far more effective strategy in the long run. Thus, the approach proposed here emphasizes the importance for less developed countries to invest in educational initiatives, campaigns or projects that build up, among their citizens, the same generative praxis that subtends the phenomenon of IT and other technological innovations of modernization presently unfolding in developed countries.

The recent reformulation of STS theory, from a focus on information technologies as material artefacts to an essentialist conceptualization of such tools as human attributes, adds further support for asserting that IT innovation practices in less advanced countries should be a self-cultivated rather than "imported" phenomenon. The design of organized environments of IT utilization should start from the principle that these advanced artefacts belong to the expressive realm of human nature. Thus, an internal capacity for engaging in improvement behaviours needs to be first devel-

---

[16] With due recognition that from a behaviorist standpoint, the lack of motivation for improvement is a functional failing of the social structure or environment, and not the individuals involved.

oped among individuals or groups, in order to ensure a fundamental *consistency* with the use of such technologies as tools or mechanisms for improving activities and processes. It is crucial to instil a psychological "ownership" of the use of IT resource capabilities among individuals and groups: in other words, the fruitful utilization of IT resources should constitute an endogenized behaviour. Since information technologies are highly useful tools for carrying out process innovation (Davenport, 1993), creating a innovation-oriented system will prepare social units and members to better assimilate with and exploit their use. It thus appears imperative to create an innate, systemic tendency towards IT innovation in societies. Recent studies bear this out (Amsden and Hikino, 1993; Perez and Soete, 1988). They show the importance of achieving an "innovating" rather than a "borrowing" strategy of growth as a way to reduce technological disparities and non-industrialization (Amsden and Hikino, 1993). It has been stated that: "a real catching up process can only be achieved through acquiring the capacity for participating in the generation and improvement of technologies, rather than in the simple use of them" (Perez and Soete, 1988, p. 459). The industries and business environments of developing/undeveloped nations need to marshal a fundamental momentum towards IT innovation behaviours, if they are to emulate the advancements of developed countries.

Recent studies on the IT productivity paradox have produced findings that claim consideration in relation to the argument raised here. Research by Dewan and Kraemer (2000, 1998), based on data of 36 countries, produced an interesting result. Industrialized countries were observed to show a significant, positive relationship between investment in IT, productivity and growth: however, there was no evidence of such a relationship in the developing countries studied. Similar results were obtained in another major study of 39 countries by Pohjola (2001). Dewan and Kramer suggested that this divergence could be explained by such factors as the comparative lack of complementary knowledge-bases, education and infrastructural resources, as well as the low level of IT investment relative to GDP, in the developing countries. While these results warrant further study, they may be taken to recommend the need emphasized here for under-developed and developing countries to focus on developing their own innovation competencies, rather than relying on loans and external consultant-led projects to increase their IT capital predominantly in the form of material assets.

4. Conclusion

The utility of conceptualizing a behavioural perspective that blends technological innovation with societal development has thus been demonstrated in terms of supporting an understanding of far-reaching transformations that can break the inertia keeping many low-income countries from enjoying a higher quality of life. In the light of the long-term benefits to a society that can accrue from learning behaviours geared towards continual technological innovation, it was asserted that low-income countries should take steps to induce structural changes to their social environments which significantly encourage the generation of innovative behaviour among members of their societies. It would do well for such countries to thus prioritize the endogenizing of IT-related innovation behaviour and competencies in their societies, over and above other policies such as encouraging heavy IT investments or pursuing rapid technical change.

The preliminary considerations raised by this paper's elaboration of the linkage between technological innovation and societal development invite further research and theoretical elaboration to develop them further. Future empirical research, for instance, might be directed towards establishing the specific nature of structural changes that have enabled particular newly-industrialized or developing countries to cultivate and mobilize practices of technological innovation, and achieve successful socio-economic growth. Nevertheless, it is hoped that this conceptual exploration will make a modest contribution to fruitful debate among researchers, and inform or challenge the views of practitioners, who are involved with the study or application of ICTs in the developing world.

References

Amsden, A. and Hikino, T. (1993) Borrowing Technology or Innovating: An Exploration of Two Paths to Industrial Development, in: Thomson, R. (ed.) *Learning and Technological Change*. London: Macmillan Press, 243–266.

Avgerou, C. and Land, F. (1992) Examining the Appropriateness of Information Technology, in: Bhatnagar, S. and Odedra, M. (eds.) *Social Implications of Computers in Developing Countries*. New Delhi: Tata McGraw-Hill Publishing, 26–41.

Avgerou, C. (1998) How Can IT Enable Economic Growth in Developing Countries?, *Information Technology for Development,* **8**. 15–28.

Blau, P. (1964) *Exchange and Power in Social Life*. New York: John Wiley and Sons.

Boyer, R. (1988) Formalizing Growth Regimes, in: Dossi, G., Freeman, C., Nelson, R., Silverberg, G. and Soete, L. (eds.) *Technical Change and Economic Theory*. London: Pinter Publishers, 608–630.

Chiaromonte, F., Dossi, G. and Orsenigo, L. (1993) Innovative Learning and Institutions in the Process of Development: On the Microfoundations of Growth Regimes, in: Thomson, R. (ed.) *Learning and Technological Change*. London: Macmillan Press, 117–149.

Chiaromonte, F. and Dossi, G. (1993) The Microfoundations of Competitiveness and their Macroeconomic Foundations, in: Foray, D. and Freeman, C. (eds.) *Technology and the Wealth of Nations: The Dynamics of Constructed Advantage*. London: Pinter Publishers, 107–144.

Davenport, T. (1993) *Process Innovation: Reengineering Work through Information Technology*. Boston: Harvard Business School Press.

David, P. (1993) Path-dependence and Predictability in Dynamic Systems with Local Network Externalities: A Paradigm for Historical Economics, in: Foray, D. and Freeman, C. (eds.) *Technology and the Wealth of Nations: The Dynamics of Constructed Advantage*. London: Pinter Publishers, 208–231.

Dewan, S. and Kraemer, K. (2000) Information Technology and Productivity: Preliminary Evidence from Country Level Data, *Management Science*. **46**, 4, 548–562.

Dewan, S. and Kraemer, K. (1998) International Dimensions of the Productivity Paradox, *Communications of the ACM,* **41**, 8, 56–62.

Dossi, G. (1988) The Nature of the Innovative Process, in: Dossi, G., Freeman, C., Nelson, R., Silverberg, G. and Soete, L. (eds.) *Technical Change and Economic Theory*. London: Pinter Publishers, 221–238.

Eijnatten, F. M. van (1993) *The Paradigm that Changed the Workplace*. Assen: Van Gorcum Publishers.

Eijnatten, F. van and Zwaan, H. van der (1998) The Dutch IOR Approach to Organizational Design: An Alternative to Business Process Re-engineering, *Human Relations,* **51**, 3, 289–318.

Emery, F. and Trist, E. (1960) Socio-technical Systems, in: Churchman, C. and Verhulst, M. (eds.) *Management Science, Models and Techniques,* vol. 2. London: Pergamon, 93–97.

Ellul, J. (1964) (tr. Wilkinson, J.) *The Technological Society*. New York: Vintage Books.

Giddens, A. (1979) *Central Problems in Social Theory: Action, Structure and Contradiction in Social Analysis*. London: Macmillan.

Giddens, A. (1984) *The Constitution of Society: Outline of the Theory of Structuration*. Berkeley, CA: University of California Press.

Goulet, D. (1971) *The Cruel Choice: A New Concept in the Theory of Development*. New York: Atheneum Publishers.

Hofstede, G. (1984) *Culture's Consequences: International Differences in Work-related Values*. Newbury Park: Sage Publications.

Homans, G. (1961) *Social Behaviour: Its Elementary Forms*. New York: Harcourt, Brace and World.

Kuhns, W. (1971) *The Post-Industrial Prophets: Interpretations of Technology*. New York: Harper and Row.

Kunkel, J. (1970) *Society and Economic Growth: A Behavioural Perspective of Social Change*. New York: Oxford University Press.

Lazonick, W. (1993) Learning and the Dynamics of International Competitive Advantage, in: Thomson, R. (ed.) *Learning and Technological Change*. London: Macmillan Press, 172–197.

Madon, S. (1993) Introducing Administrative Reform through the Application of Computer-Based Systems: A Case Study in India, *Public Administration and Development,* **13**, 1, 37–48.

Matthews, J. (1997) Introduction to the Special Issue, *Human Relations,* **50**, 5, 487–496.

McClelland, D. (1961) *The Achieving Society*. Princeton: Van Nostrand Reinhold.

Mokyr, J. (1990) *The Lever of Riches: Technological Creativity and Economic Progress*. New York: Oxford University Press.

Mumford, E. (1996) *Systems Design: Ethical Tools for Ethical Change*. Basingstoke: Macmillan.

Nelson, R. (1993) Technical Change as Cultural Evolution, in: Thomson, R. (ed.) *Learning and Technological Change*. London: Macmillan Press, 9–23.

Odedra-Straub, M. (ed.) (1996) *Global Information Technology and Socio-economic Development*. Nashua: Ivy League Publishing.

Orlikowski, W. (1996) Learning from Notes: Organizational Issues in Groupware, in: Kling, R. (ed.) *Computerization and Controversy: Value Conflicts and Social Choices,* (2nd ed.), San Diego: Academic Press, 173–189.

Owens, E. (1987) *The Future of Freedom in the Developing World: Economic Development as Political Reform.* New York: Pergamon Press.

Pasmore, W. (1995) Social Science Transformed: The Socio-Technical Perspective, *Human Relations,* **48,** 1, 1–22.

Pasmore, W. (1988) *Designing Effective Organizations: The Sociotechnical Systems Perspective.* New York: John Wiley & Sons.

Perez, C. and Soete, L. (1988) Catching up in Technology: Entry Barriers and Windows of Opportunity, in: Dossi, G., Freeman, C., Nelson, R., Silverberg, G. and Soete, L. (eds.) *Technical Change and Economic Theory.* London: Pinter Publishers, 458–479.

Pohjola, M. (2001) Information Technology and Economic Growth: A Cross-country Analysis, in: Pohjola, M (ed.). *Information Technology and Economic Development.* Cambridge, MA: Oxford University Press, 242–256.

Porter, M. (1990) *The Competitive Advantage of Nations.* New York: Free Press.

Quah, D. (2000) Valuation and the New Economy, *Report for Merrill Lynch,* 9th May. econ.lse.ac.uk/staff/dquah/p/0006ml.pdf.

Rapley, J. (1996) *Understanding Development: Theory and Practice in the Third World. Boulder.* Lynne Rienner Publishers.

Rosenberg, N. (1994) *Exploring the Black Box: Technology, Economics and History.* New York: Cambridge University Press.

Rosenberg, N. (1982) *Inside the Black Box: Technology and Economics.* New York: Cambridge University Press.

Sitter, L. de, Hertog, J. den and Dankbaar, B. (1997) From Complex Organizations with Simple Jobs to Simple Organizations with Complex Jobs, *Human Relations,* **50**, 5, 497–534.

Smith, K. (1982) Philosophical Problems in Thinking about Organization Change, in: Goodman, P. (ed.) *Change in Organization.* San Francisco: Jossey Bass, 316–374.

Smith, J. (1993) Thinking about Technological Change: Linear and Evolutionary Models, in: Thomson, R. (ed.) *Learning and Technological Change.* London: Macmillan Press, 65–79.

Talero, E. and Gaudette, P. (1995) Harnessing Information for Development: A Proposal for a World Bank Group Vision and Strategy, *Information Technology for Development,* **6**, 145–188.

Thomson, R. (1993a) Introduction, in: Thomson, R. (ed.) *Learning and Technological Change.* London: Macmillan Press, 1–5.

Thomson, R. (ed.) (1993b) *Learning and Technological Change.* London: Macmillan.

Todaro, M. (1994) *Economic Development,* (5th ed.). New York: Longman Publishing.

About the Author

Stephen Corea received his PhD recently from the London School of Economics, and is currently a lecturer in Information Systems at Warwick Business School, UK. His research interests centre on the areas of IT innovation cum organizational transformation, IS strategy, and IT governance, in which he is presently engaged on the development of sociotechnical perspectives and systems thinking. His background in industry includes previous positions as an IS manager, systems analyst and network administrator.

3. Context-Specific Rationality in Information Systems

Chrisanthi Avgerou

London School of Economics
c.avgerou@lse.ac.uk

Abstract

Information Systems (IS) studies are primarily concerned with the construction of new ICT-based artefacts and their take-up within organizations—mainly business firms. Most of the IS literature is based on the assumption that processes of technology and organizational innovation are driven by a mix of three rationalities, that is, the following three modes of reasoning—techno-scientific reasoning of engineering; managerial reasoning for effective organizing in terms of competitive positioning a business firm in a market context; and economic reasoning for optimizing economic utility for the individual and the firm.

In this chapter I challenge the assumption that IS practice can be understood as employing these types of universal rationality. I point out that the literature of development studies, and IS in developing countries in particular, suggests that the combination of rationalities that prevails in current IS and organizational development knowledge is ineffective in mobilizing local human capabilities to improve socio-economic conditions.

Consequently, I argue for a context-specific notion of rationality for the study of information systems. Drawing from Foucault's concept of knowledge as being inseparable from power relations within society, I introduce the concept of "organizing regime" to refer to historically constructed rationalities within which IS innovation is embedded.

1. Introduction

From the outset, information and communication technology (ICT) has been a triumph of rational thinking of Western modernity—the enlightenment ideal of making progress through reason fuelled the continuous scientific and technological advances that led to the sophisticated modern computers and telecommunications. Furthermore, the diffusion of ICT innovation that created an unprecedented infrastructure of data and communication channels within and among organizations across the globe took place in societies apparently committed to create wealth by optimizing efficiency. These two rational courses of action—scientific advancement and the quest for economic efficiency and growth—are not only compatible, but they re-enforce each other in a virtuous relationship—technology is one of the most significant variables in economic growth, and the economically strong societies invest in continuing research and development.

The view that socio-economic improvement is a result of technical/rational action is a dominant premise for information systems practice and research. The bulk of the knowledge developed in two major sub-fields of information systems, systems development and management information systems, assumes that technology is deployed in the context of an enterprise striving for competitiveness in a free market economy.

In a nutshell, three "pure" types of rationality are intermixed in information systems studies—the engineering rationality, the administrative rationality and the economic rationality. We can think of rationality as calculative behaviour that focuses on a particular concern, or aims at achieving a desirable end. Invariably, the study of calculative behaviour involves the analysis of reasons for actions and the identification of logically consistent courses of action. Many academic disciplines tend to express notions of rational action in formal models, theories, and formal notations. These are necessarily abstractions, often based on unrealistic assumptions about the circumstances of individuals and organizations.

The problem is that social phenomena, such as the introduction of new technologies in organizations, generally involve multiple and often conflicting concerns. Moreover, it is frequently the case that these social phenomena are vaguely expressed and only partially understood by organizational actors. Not only does the intertwining of the three pure rationalities increase the complexity of IS innovation, rendering reductionist technical analyses inadequate to capture the unfolding dynamics of such action, but all too often actors' behaviour diverts from what is "rational" according to the combination of engineering, administrative and economic logic.

Such diversions of rational techno-economic courses of innovation in organizations are clearly manifested in developing countries. In a history of about 60 years of frustrated attempts to foster economic growth by imitating the rationalities that have underpinned the institutions of the rich Western societies, modernization in developing countries has proved a difficult and often destructive process. ICT is now entering developing countries as the most significant force of modernization (Kirkman *et al.*, 2002). However, there are clear signs that professional IS practice fails to deliver the effects from ICT to which we are accustomed, given our experience of its wide diffusion in industrialized countries (Heeks, 2002). Many projects are resisted, although the administrative and economic benefits they are aiming at may be generally desirable. The technical integrity of professional activities is often violated by apparently irrational interferences. Even when technology systems are successfully implemented, their use may drift and be assimilated by the very mode of inefficient and ineffective functioning of the organization that they were intended to change.

The general tendency in the IS literature is to see such cases as manifestations of irrationality and either to insist on the "correct" application of technical/rational courses of professional knowledge by engineers and managers, or to attempt to develop "soft" methods and remedial action for a feasible transition from an "irrational"/traditional situation, to the rational/modern organizational condition.

In this chapter, I question this underlying premise of ICT and modernization and argue that the intertwined engineering, administrative and economic rationalities that have been a powerful engine for Western modernity do not constitute a universally valid logic in the way the laws of nature are understood to be. Section 2 briefly explains the engineering, administrative and economic rationalities and section 3 explores the meaning of rationality, mainly drawing on Weber's (1947) analysis. Section 4 presents a historically constructed view of rationality, drawing on Foucault's (1977, 1980) association of power and knowledge and his notion of a "regime of truth". Then, section 5 examines how this critique applies to the idea that life conditions in developing countries can improve by the ideology of modernization, followed by section 6, which introduces the notion of "organizing regime" to refer to the historically formed regimes of truth of organizations. Finally, the conclusion states the core arguments of this chapter, i.e., that IS innovation should be seen as being situated at the junction of the local organizing regimes, universal logic conveyed by ICT products and professional experts, and international institutions that constitute the current condition of globalization.

2. The Pure Rationalities of Information Systems Study and Practice

In the typical fashion of the modern academic analytical thinking, the study and practice of ICT in organizations has tended to reduce the complexity of the phenomena of IS innovation and address them in terms of one predominant effort, namely in the case of IS—to construct technology artefacts; to accommodate technical artefacts in organizations; and to harness technology and organizational innovation for economic benefits. For each of these efforts, there is a tradition of analytical thinking, an elaborate way of reasoning, a "rationality".

The engineering rationality is concerned with the construction of reliable technology artefacts. This concern is captured in models, theories, or good professional practice recommendations with varying degrees of formality. At one extreme, there are formal methods in software engineering, which present the whole process of software development as a succession of operations underpinned by formal logic and mathematics. The formal engineering rationality is in most cases of software engineering practices unrealistic. There are no readily available formal models that capture the behaviour of the target system, such as the information processing involved in the day to day actions that constitute the processes of a business firm or a hospital. However, softer versions of the rational engineering model are widespread. The information systems—"life cycle", which has been the most fundamental concept in IS training and practice has obvious engineering underpinnings. Moreover, the various methods and methodologies for systems development that have been worked out for the realization of the life cycle are common professional tools.

The administrative rationality is concerned with effective organizing for business and communal service purposes. Max Weber's (1947) analysis of ideal types of organizations, and the determination of one particular form of hierarchical organization, the bureaucracy, as the most prevalent way of thinking about and managing organizations in the twentieth century, has been extremely influential (Albrow, 1970). The realization of the bureaucracy model has faced serious undesirable complications and came to be seen as deeply problematic rather than ideal. Nevertheless, many of its principles, such as the hierarchical lines of control and specialized roles for the performance of functional duties still remain among the most fundamental ways in which we think about organizations. Other ways of reasoning about effective organizing include the analysis of decision making and operational research. Each of these ways of thinking about organizing has its own internal integrity and contributes techniques, which form part of the expertise of various management roles.

The economic rationality is concerned with maximizing the economic benefit of the individual, the organization, or a larger social unit such as a country. The belief that the actions of individuals and consequently the organizations they form are governed primarily by their concern to maximize economic benefits is one of the main characteristics of modern Western society. Not surprisingly there has been a great deal of interest in analyzing ICT innovation in organizations as an economic maximizing endeavour, i.e., in terms of economic benefits, costs, and risks. A specialist stream of information systems economics is devoted to the analytical understanding of the economic gains and losses implicated in ICT innovation. But the practice of such technical economic reasoning remains limited. It is widely known that the forecast and evaluation of economic benefits, costs, and risks are the least performed tasks in professional practice. One of the most enduring issues in the IS literature is that information systems projects are launched without careful economic analysis and completed projects are rarely assessed for their economic value.

Although these rationalities are rarely found applied in their pure form, they are more frequently discernible in professional practice in combination with each other. There are many examples of IS models, methods and good practice recommendations which serve mixed concerns of engineering, organizing, and economic gain. The more widely used versions of the systems life cycle model try to do this with a variety of techniques. In addition to producing technically valid specifications, for example with object oriented analysis and design techniques, they also guide professional attention to improving—or radically changing—organizational processes and structures, as in the requirement analysis stage, and they always recommend some form of economic evaluation, as a kind of investment appraisal. Sometimes the engineering and organizational concerns are combined more explicitly, as in Mumford's ETHICS (Mumford and Weir, 1979) methodology or more recently in the various methodologies of ERP implementation (Markus *et al.*, 2000). In IS management, the economic and organizational rationalities are intertwined so closely that they are almost inseparable, see for example Porter and Millar's (1984) model analyzing the role of IT for business competitiveness. Ultimately, as the Business Process Re-engineering literature of the 1990s indicated (Hammer and Champy, 1993), IS innovation comes to be confronted with the combination of all three rationalities as it is expected to achieve economic gains in the free market economy through organizational transformation supported by the design and implementation of new ICT artefacts.

However, the main issue of interest to me here is not whether such a combination of technical/rational courses of action is generally feasible to pursue by professionals in modern organizations or effective. Despite the concerns about frequent IS project failures (Sauer, 1999), there is enough evidence that such complex action is constantly pursued by competent professionals in the corporations of the advanced economies, in which ICT innovation and transformations of business practice and structure contribute to their global competitive advantage. Rather, the main problem I am interested in addressing regarding ICT in contemporary social and organizational change is that such a combination of rationalities proves difficult to follow successfully in most organizational and social settings that are not already accustomed to the business management tradition of the Western advanced free market economies. There are conspicuous differences in the "success" of ICT innovation both among different sectors of societies (with public sector services and small enterprises struggling to adopt best practices and to achieve expected rationalization results in their performance) and among societies themselves (with developing countries struggling to accommodate IS innovation in their organizations) (Avgerou, 2000).

3. Formal vs. Substantive Rationality

One of the most important theoretical efforts to address rationality in the context of Western modernity is found in the work of Max Weber. Written in the beginning of the twentieth century, while the course of modernity was already quite advanced,

Weber attempted to theorize the "rational" consistency of fundamental institutions of modernity, such as the order that sustains organizations and the functioning of the free market economy.

Weber's main notion of rationality, "formal rationality", is primarily a methodological device for the purposes of typological scientific analysis. His methodology consisted of constructing ideal categories of social settings and determining a course of rational action for the purposes of their particular ends. Such an abstract "purely rational" course of action may then serve as a basis of comparison with actual, observed action to account for deviations from the line of conduct which would be expected given the hypothesis that the action was purely rational. Within this framework of analysis, a particular behaviour or action is seen as irrational if it deviates from the conceptually pure type of rational action. Action which does not conform to the rational means of serving the end assumed by the ideal model of action, is irrational.

Seen in such a comparative perspective, the modern Western free market economic order is not a "natural order", but one possible line of social development. Weber himself identified other rationalities built into the dominant institutional structures of different societies, such as the "communal rationality", referring basically to the family-based economic structures of traditional societies.

Weber theorized also on the social conditions that support particular rationalities. One of the most significant contributions of his analysis of the rational economic activity under market conditions is the outline of the fundamental social conditions of the modern capitalist system. These include market freedom, autonomy in the selection of management by the owners, free labour market, absence of regulation of consumption, production, and prices, or of other forms of regulation, calculability of the technical conditions of the production process, a public administration legal order, complete separation of the enterprise from the household or private budgetary unit and its property interests, and a formally rational monetary system (Weber 1947).

In contrast, in societies oriented to a communal rationality, economic activity is neither clearly differentiated from other action, nor does it involve the formal calculated arrangements of "capital accounting". Elements of reciprocative and redistributive rationality such as mutual responsibility among members of the community, often the extended family, solidarity, and the common welfare of members prevail over the freedom of enterprise to accumulate capital. Moreover, affectual, emotional and traditional ties are legitimate motives for action.

In analyzing the formal rational type of the modern economy, Weber acknowledges that the outcome of economic action is judged differently in relation to different underlying ends, which he calls "substantive rationality". Such ends may include ethical, political or utilitarian considerations, such as social equity, social justice, furtherance of power of a political unit. For example, a particular course of economic action may be successful in terms of social equity within a social group, but inadequate in terms of the overall power of the social group *vis-à-vis* its political rivals. Indeed, in terms of substantive rationality, economic activity itself may be of secondary importance, or in conflict with the attainment of particular social values of a society.

The notion of the substantive rationality provides a wider perspective to view the modern economic system. Rational economic action describes what course human action should take to be in accordance with an ideal model if it were completely and unequivocally directed to a single end: the maximization of economic advantage. But such a "rationalized" economy, where people orient their decisions towards maximizing efficiency and weighing costs and benefits, is based on a particular mentality, involving the ethical sanction of acquisitive activity and a propensity to seek new solutions to problems rather than to adhere to traditions.

In this sense, the modern Western economic rationality that values productivity and efficiency in a free market setting conveys one particular set of values that has dominated over others. This thesis has been echoed in many other studies that look at the economic activity within a social system. For example, Parson and Smelser (1956) made the point that societies differ with respect to the degree to which they temper economic productivity and efficiency in relation to other values. As a result, in different historical settings different mixes of economic and non-economic rationalities prevail and are perpetuated through institutional mechanisms. Thus, Smelser (1978) suggests the need to understand the way different kinds of rationalities that may govern the production, distribution, and consumption of economic goods and services, such as efficiency, social justice, social security and military defence, are incorporated in the complex economic and social processes of particular societies, either local or global.

4. Critique of the Idea of Rationality

Although based on the relativistic premise that different rational systems of ideas and activities may be valid for different groups of people in different social settings, Weber's work has contributed to the legitimization of the supremacy of the institutions of modernity rather than challenging them. The technical distinction between rational and irrational action proved problematic. It overlooks the substantive sense of rationality, that is also pointed out in Weber's analysis. A deviation from a course of action dictated by the rational norms of a formal ideal type of social system may be rational in serving a different end from that assumed in this particular formal abstraction.

Weber considered it a danger for the social sciences to pursue rationalistic interpretations, i.e., to assume the predominance or general desirability of the courses of action he determined in his abstract ideal rational categories. Yet, his formal rational models had an influence beyond academic analysis, inspiring, guiding and legitimating patterns of social action. His ideal types of authority, for example, which define and compare the charismatic, traditional and rational-legal modes of commanding and maintaining order, and distributing resources, contributed to the making of the bureaucratic model of organization the most widely spread institution of the modern western society.

Since Weber's time, a succession of theoretical perspectives have surpassed his analysis, taking a critical stance to the rationality of modernity. They include the Marxist critical theorists Adorno, Horkheimer (Horkheimer 1947; Horkheimer and Adorno 1972), and Marcuse (Marcuse 1964); Foucault (Foucault 1977); the various theorists of post-modernism, such as Lyotard (Lyotard 1984); the social-constructionists (Hughes, 1983; Latour and Woolgar, 1986; Callon and Law, 1989; Bijker and Law, 1992; Latour, 1993); and sociologists of the "late modernity" such as Giddens, Beck, and Lash (Beck *et al.*, 1994).

It is beyond the scope of this chapter to present and discuss these paths of critical thinking. Instead, I will take a closer look at a particular perspective on the nature of rationality based on the work of Michel Foucault (1980). For Foucault, rationality is a constituent part of the ubiquitous politics of determining what is true and what is false. Fundamental in Foucault's analysis is the conception that knowledge and power are constituted in interdependence to each other; his studies trace the history of how particular types of discourse in particular domains such as sexuality, criminality and madness, came to be considered rational. He uses the term "regime of truth" to capture the socially constructed, power constituted determination of what is rational—in other words of what is a valid way to distinguish between true and false. Regimes of truth are the rules according to which truth is determined and specific effects of power are attached to the true.

Foucault does not examine the intrinsic rationality of scientific and economic practices, their overall capacity to determine what is true in nature and social affairs, and their general effects on the human condition. He does not try to understand whether certain general kinds of rationality are a fundamental mechanism for the organizational and institutional order, or represent repression, coercion and violence, and therefore form a mechanism for social domination. Instead he considers particular discourses shaped under particular political, economic, institutional regimes of the production of truth. Fundamental in his approach is a conceptualization of power as something which cannot be *a priori* judged either as positive or negative.

Foucault's studies simply implicate power in the formation of regimes of truth, which he considers essential for the structure and functioning of societies. Power is present in all discourses, interwoven with all kinds of relations, such as kinship, sexuality or production, and in multiple forms, either negative such as prohibition or positive such as protection. General conditions of domination are created by such multiform localized power-laden relations integrated into overall strategies. But, relations of power are always accompanied by resistances, which are also multiple in form and can be integrated to form larger strategies. Indeed, Foucault's historical analyses of rationality attempt to uncover the "knowledges" that, in the struggles and conflicts of the formation of particular social institutions, were disqualified as inadequate and naive, what he calls the "subjugated knowledges". Thus he pays attention to forms of domination as well as to the resistances encountered.

This is where the radical nature of Foucault's work lies. By uncovering the subjugated knowledges in the history of the formation of prevalent modern social institutions and by tracing the omnipresent effects of power that condition reason, he strips the most well respected domains of knowledge and practice, such as those on which medical and penal systems are based, of their rationality privilege.

5. The Rationality of Modernization as "Development"

The concepts of rationality, as well as Foucoult's notion of the regime of truth outlined above, have been proposed to account for the social context of industrialized western societies, but they have clear implications for the so-called developing countries. If the techno-economic rationality of modernity is not a neutral and objective way of reasoning but is, instead, an aspect of the culture of western civilization, and thus inextricably bound to its history, the changes brought by modernization into the life conditions in the poorer and politically weaker require a closer examination. There have been many voices critical of the way the universalist rationality of western modernity in its various forms—colonialism, economic modernization, socialist totalitarianism—has affected the social fabric of communities around the world, see for example (Freire, 1970; Nandy, 1987; Said, 1994; Gandhi, 1997). These voices juxtapose the local and traditional logics of the sacred, the symbiosis with nature and self-sufficiency, with the failure of the reason of efficiency, growth and science to sustain a satisfactory state of human life. Here I will only review briefly Escobar's (1995) critique of the development interventions, which draws directly from the western critiques of the rationality of modernity reviewed above.

Escobar (1995), following the critical line of Foucault, analyzed the notion and policies of socio-economic development pursued since the 1940s as a discourse, that is as a space of thought and action within which only certain things can be said, done, or imagined. Examining the political dynamics, theoretical ideas and the practical interventions that have constituted development, he argues that the adoption in developing countries of free market rationality and its institutions, as well as the cognitive instruments of science and technology, was a socially constructed adoption rather than being naturally chosen.

The organizing premise of the discourse of development has been the belief that modernization, based on the two pillars of science-technology and capital, is the only force capable of destroying archaic superstitions and relations, and that it should be applied at what ever social, cultural and political cost. Within this discourse, institutions and professionals of development have determined and classified problems by applying the concepts and techniques of the sciences that sustain modernity—economics, public administration and management—and formed policies of change.

Presented as a detached rationality capable of improving the human condition, the modernization discourse has created a regime of truth, passing judgement on social groups, determining their needs, and prescribing how they should change. Social life has been conceived as a technical issue, and its improvement is entrusted to technical experts, capable of rational decision making and management.

Escobar's analysis points out that, instead of delivering universal improvement of the human condition as was initially expected, indiscriminate application of instrumental rationality in the second half of the twentieth century eroded poor people's ability to define and take care of their own lives to an even greater extent than the erosion of past colonial regimes. The transfer of the rationality of modernity carried with it the transfer of values and institutions. For example, the need for foreign exchange and investment influenced the promotion of cash crops to the detriment of food crops for domestic consumption; targeting efficiency and competitiveness in the global economy imposed industrialization—or post-industrialization—interventions over local production and trade patterns.

Moreover, Escobar noted, a discourse that privileges a modern culture of western values and modes of knowledge and action is destructive in a deeper manner. It considers local cultures, predominant values and politics responsible for backwardness. Most interventions of modernization have paid little attention to the historically derived system of values that sustain social systems—such as an economy, a business organization, a public service institution—which are irrational from the point of view of the rationality of modernity.

6. Organizing Regimes

Let us now turn to examine the rationality assumptions made in organizational practice and information systems innovation. We can see the pure engineering, administrative and economic rationalities in Weberian terms as ideal types of behaviour, abstractions of complex processes taking place in the context of modern societies.

Indeed, as mentioned above, most studies of organizations and information systems moderate the assumption of universalist techno-economic rationality, and a stream of research has elaborated on the messiness of organizational behaviour. Herbert Simons notion of bounded rationality (Simon 1945) set the tone for recognizing the limitations of organizational reasoning even in technical/rational studies. Others studied the way social, cultural, and political behaviours complicate the fundamental techno-scientific rationality, without, however, always challenging it as the underlying principle of organizing. This is the case with many studies, which reveal such complications and moderate, but do not abandon, the technical/rational behaviour thesis. They still "appear to regard the spread of 'western' norms of rationality as almost inevitable given the development of world markets, ... (and) do not discuss whether there may be alternative variants of such rationality" (Whitley 1997, p. 290).

Nevertheless, there have been studies that highlighted examples of organizations in which individual and collective action clearly does not comply with the assumption of such principles of rational behaviour and which exhibit what Clegg calls "alternative modes of rationality" (Clegg 1990). The organizational behaviour patterns in some East Asian countries, mainly Japan, South Korea and Taiwan, have been repeatedly discussed as examples of alternative underlying substantive rationalities, see for example (Hamilton and Biggart 1988; Clegg 1990). It has been pointed out that the organization of Japanese industry in groups of firms is based on a communitarian ideal, which involves the consideration of what is good for the collectivity, not for individual firms. South Korean conglomerates are an expression of a patrimonial principle, whereby an authoritarian leader acts in the mode of a patriarch and his children. Taiwanese groups of firms are based on familial relations which, however, are not under the control of a single patriarch. Many other examples of alternative modes of rationality can be found in studies of organizations in Latin America or Southern Europe (Clegg 1990; Kumar and van Dissel 1998; Avgerou 2002).

Generally in such studies, alternative modes of rationality are seen as being derived from long-term broader social experiences within which the organizations under study are embedded. For example, differences in the degree to which rational action is driven by trust and co-operation, or competition and antagonism, are traced in broader, historically formed, structures of social relations (Fox 1974; Dore 1983; Gambetta 1988).

The recognition of multiple, historically developed, substantive rationalities and congruent modes of organizing provides an orientation towards a contextualist perspective in the study of IS innovation (Avgerou 2001). Rather than aiming to develop general knowledge and practice to conform to a-contextual rules of rational behaviour, the recognition of differences in substantive rationality directs attention to local meanings and legitimate action. However, with all the importance of an epistemology that overcomes the rigidity of a universalist conception of rationality, associating modes of rationality with the historically developed social relations of a locality, whether a country or an otherwise defined region, is a crude thesis for three reasons. First, it entails the risk of stereotyping and oversimplifies the multiple influences exerted on organizations in the current condition of globalization. Second, emphasizing the difference among underlying substantive rationalities across local communities creates the impression of homogeneous beliefs and behaviours, overlooking the co-existence of multiple beliefs, in particular ignoring the subjugated beliefs and knowledges. In order to overcome such misleading oversimplifications of rationality and context, we need to examine more closely the way an individual's behaviour is related to historically shaped organizing regimes rather than universalist types of rationality; Weick's concept of enactment (Weick 1979) provides useful insights to that end.

According to the notion of enactment, individuals and groups are not passive recipients of influences from their institutional context: they "construct, rearrange, single out, and demolish many 'objective' features of their surroundings" (Weick, 1979, p. 164). However, individuals' interaction with their organizational context by enactment is very different from the notion of action pre-calculated to produce efficiency benefits. In enactment, members deal with the "here and now", according to their desired outcomes which stem from their beliefs, according to their perceptions of entailed gains or risks. These perception and beliefs are neither arbitrary nor reducible to universal fundamentals of human nature or social history; they are derived from life conditions, embodied life experiences (cf. Verran 1999). There is therefore a continuous interaction between the behaviour of subjective rational actors and social institutions. Individuals' and groups' actions form, sustain and change the institutions in which they are embedded and which subsequently feed into their perception of resources and constraints afforded to their action. This understanding of rationality in an organization's context is taken in a number of IS studies, most prominently by Ciborra (1991, 1999), Orlikowski (1996) and Suchman (1987).

Organizing entails the steering of individuals' enactments in collective action. The achievement of such a congruence of subjective situated behaviour is partly a matter of institutional influences of the environment, either by the shaping of substantive rationalities or by coercion, and partly internal institutionalization processes, again a matter of culture and/or coercion. A point that needs to be emphasized is that organizing is subject to multiple external influences and multiple internal enactments; therefore it cannot be seen as the result of common perceptions, common values and desired outcomes, and a single knowledge. Rather, organizing entails a regime of truth in Foucault's terms, with multiple knowledges streamlined in particular enactment co-ordinations under historically formed power relations. Paraphrasing this sentence, the rationality of an organization is a particular organizing regime.

It is important to point out that the rationality of organizing behaviour is shaped by influences stemming from multiple sources beyond the organization's locality— whether this is understood as the country or the region. More importantly, social institutions are not culturally, socially, and politically harmonious and homogeneous. There too, manifested dominant rationalities hide subjugated knowledges. To wit, an organizing regime is woven from multiple enactments and social/historical influences. While particular rationalities are dominant in shaping the overall consistency of collective organizational behaviour, other substantive rationalities, other perceptions of truth, are still present, even if not engaging in active resistance.

7. Conclusions

An information systems project may strengthen or challenge the established conditions of power/knowledge, that is, the existing organizing regime. Even when it expresses the dominant rationality of an organization, the setting in motion of an information systems innovation project requires the collaboration of others—the users, the customers, the various categories of professionals involved in the established organizing practices. Being shaped by their histories in different social settings, each of these categories of actors may have different material concerns, different understandings of the consequences of innovation that matter, different symbolic expressions of what they know and what they want.

From such a perspective, the mainstream knowledge on ICT use of information systems professionals and the rationality of efficiency and competitiveness driving business-school-trained management professionals can be seen as constituting a widespread institutionalized regime of truth. This organizing regime is constantly contested, and the results of the innovation depend on the outcome of many situated confrontations with subjugated knowledges. In order to account for the diversity of outcomes of IS innovation, we need to be able to account for the alternative knowledges of the actors on whose enactment the information systems innovation relies, i.e., the alternative logics that are rooted in the substantive rationalities of the way they have lived their lives.

In other words, there can be no general *a priori* dichotomy between rational and irrational, rational and ideological, true and false, success and failure. In every arena of the multiple contested rationalities of the actors' institutionalized conditions, some rationalities have more legitimacy than others. It is important to note that the techno-economic rationality of ICT and management is more compatible with certain organizations' organizing regimes and more alien with others. This is a general problem in the so-called technology transfer cases in developing countries, in which technologies and organizational practices that are shaped and legitimized in one social setting are introduced in another. From the perspective of multiple situated rationalities, it is not surprising that the literature of development studies is littered with accounts of the friction that takes place in the receiving organizations.

References

Albrow, M. (1970) *Bureaucracy*. London: Macmillan.

Avgerou, C. (2001) The Significance of Context in Information Systems and Organizational Change, *Information Systems Journal*, **11**, 1, 43–64.

Avgerou, C. (2002) *Information Systems and Global Diversity*. Oxford: Oxford University Press.

Beck, U., Giddens, A. and Lash, S. (1994) *Reflexive Modernization*. London: Macmillan.

Bijker, W. E. and Law, J. (eds.) (1992) *Shaping Technology / Building Society*. Cambridge, Massachusetts: The MIT Press.

Callon, M. and Law, J. (1989) On the Construction of Sociotechnical Networks: Content and Context Revisited, *Knowledge and Society,* **9**, 57–83.

Ciborra, C. U. (1991) From Thinking to Tinkering: The Grassroots of Strategic Information Systems, *Proceedings of the 12th International Conference on Information Systems*, New York: 283–292.

Ciborra, C. U. (1999) A Theory of Information Systems Based on Improvisation, in: Currie, W.L. and Galliers, R. (eds.) *Rethinking Management Information Systems*. Oxford: Oxford University Press: 136–155.

Clegg, S. R. (1990) *Modern Organizations: Organization Studies in the Post-modern World*. London: Sage.

Dore, R. (1983) Goodwill and the Spirit of Market Capitalism, *British Journal of Sociology*, **34**, 459–482.

Escobar, A. (1995) *Encountering Development*. Princeton: Princeton University Press.

Foucault, M. (1977) *Power/Knowledge*. New York: Pantheon.

Foucault, M. (ed.) (1980) *Power/Knowledge: Selected Interviews and Other Writings 1972–1977*. New York: Prentice Hall.

Fox, A. (1974) *Beyond Contract: Work, Power and Trust Relations*. London: Faber and Faber.

Freire, P. (1970) *Pedagogy of the Oppressed*. New York: Seabury Press.

Gambetta, D. (ed.) (1988) *Trust: Making and Breaking Co-operative Relations*. Oxford: Blackwell.

Gandhi, M. (1997) The Quest for Simplicity: My Idea of Swaraj, in: Rahnema, M. and Bawtree, V. (eds.) *The Post-Development Reader*. London: Zed books: 306–307.

Hammer, M. and Champy, J. (1993) *Reengineering the Corporation: A Manifesto for Business Revolution*. London: Nicholas Brealey.

Heeks, R. (2002) Information Systems and Developing Countries: Failure, Success and Local Improvisations, *The Information Society*, **18**, 2, 101–112.

Horkheimer, M. and Adorno, T.W. (1972) *Dialectic of Enlightenment*. New York: Herder and Herder.

Horkheimer, M. (1947) *Eclipse of Reason*. New York: Continuum.

Hughes, T.P. (1983) *Networks of Power: Electrification in Western Society 1800–1930*. Baltimore, MD: Johns Hopkins University Press.

Kirkman, G. S., Sachs, J., Schwab, K. and Cornelius, P. K. (2002) *The Global Information Technology Report 2001–2002: Readiness for the Networked World*. New York: Oxford University Press.

Kumar, K. and Dissel, H.G. van (1998) The Merchant of Prato—Revisited: Toward a Third Rationality of Information Systems, *MIS Quarterly*, **22**, 2, 199–226.

Latour, B. (1993) *We Have Never Been Modern*. New York: Harvester Wheatsheaf.

Latour, B. and Woolgar, S. (1986) *Laboratory Life: The Constitution of Scientific Facts*. Princeton: Princeton University Press.

Lyotard, J. F. (1984) *The Post-Modern Condition: A Report on Knowledge*. Manchester: Manchester University Press.

Marcuse, H. (1964) *One-Dimensional Man*. Boston: Beacon Press.

Markus, M. L., Axline, S., Petrie, D. and Tanis, C. (2000) Learning from Adopters' Experiences with ERP: Problems Encountered and Success Achieved, *Journal of Information Technology*, **15**, 4, 245–265.

Mumford, E. and Weir, M. (1979) *Computer Systems in Work Design: The ETHICS Method*. London: Associated Business Press.

Nandy, A. (1987) *The Intimate Enemy*. Bombay: Oxford University Press.

Orlikowski, W. J. (1996) Improvising Organizational Transformation over Time: A Situated Change Perspective, *Information Systems Research*, **7**, 1, 63–92.

Parsons, T. and Smelser, N. J. (1956) *Economy and Society*. New York: The Free Press.

Porter, M. and Millar, V. (1984) How Information Gives you Competitive Advantage, *Harvard Business Review*, **63**, 4, 149–160.

Said, E. (1994) *Culture and Imperialism*. New York: Vintage Books.

Sauer, C. (1999) Deciding the Future for IS Failures: Not the Choice you Might Think, in: Galliers, R. and Currie, W. L. (eds.) *Rethinking Management Information Systems*. Oxford: Oxford University Press: 279–309.

Simon, H. (1945) *Administrative Behavior*. New York: Macmillan.

Smelser, N. J. (1978) Re-examining the Parameters of Economic Activity, in: Epstein, E. M. and Votaw, D. (eds.) *Rationality, Legitimacy, Responsibility: Search for New Directions in Business and Society*. Santa Monica: Goodyear Publishing Company: 19–51.

Suchman, L. (1987) *Plans and Situated Action*. Cambridge: Cambridge University Press.

Verran, H. (1999) Staying True to the Laughter in Nigerian Classrooms, in: Law, J. and Hassard, J. (eds.) *Actor Network Theory and After*. Oxford: Blackwell: 136–155.

Weber, M. (1947) Sociological Categories of Economic Action, in: Parsons, T. (ed.) *Max Weber: The Theory of Social and Economic Organization*. New York: The Free Press: 158–323.

Weick, K. (1979) *The Social Psychology of Organizing*. New York: McGraw-Hill.

Whitley, R. (1997) The Institutionalist Approach, Review of the 'Institutional Environments and Organizations: Structural Complexity and Individualism', by Scott, W. R. and Meyer, J. W. (eds.), *Organization*, **4**, 2, 279–302.

About the Author

Chrisanthi Avgerou is Professor of Information Systems at the London School of Economics and Political Science (UK). She is chairperson of the IFIP technical committee 9 on "Computers and Society" and she chaired the IFIP WG 9.4 on "Social Implications of Computers in Developing Countries" from 1996–2003. She teaches postgraduate courses on Information Systems Implementation and on Information Systems in Developing Countries. Her research is concerned with the study of the dual process of the utilization of information technology and organizational change within different socio-organizational contexts. Among her recent publications is *Information Systems and Global Diversity* (2002), published by Oxford University Press.

4. Support Environments for E-Learning in Developing Countries

Noorminshah Iahad

Interactive Systems Design Group
Computation Department
UMIST, UK

n.iahad@postgrad.umist.ac.uk

Georgios A. Dafoulas

Business Information Systems Group
School of Computing Science
Middlesex University, UK

g.dafoulas@mdx.ac.uk

Interactive Systems Design Group
Computation Department
UMIST, UK

dafoulas@co.umist.ac.uk

Abstract

For developing countries to participate actively in the global knowledge-economy, it is essential that their people acquire relevant skills and knowledge. E-Learning, simply defined as "learning through the Internet", has been identified as an efficient knowledge transfer mechanism. It is one of many applications that has evolved from the broader E-Commerce concept. Specifically, it is one of the e-services provided to consumers in business-to-consumer (B2C) E-Commerce transactions, which concern knowledge dissemination.

The successful implementation of an E-Learning environment depends on the development of the appropriate infrastructure and the support environment. Problems such as an inadequate number of personal computers per household, poor Internet connectivity, and insufficient bandwidth need to be resolved before an appropriate support environment for E-Learning can be created. Such an environment must of necessity ensure the relevance and quality of the Content, Assessment, Course Management and Communication. The environment must not only consider technology-related aspects but also pedagogical aspects of E-Learning. This will ensure that the demand for a learner-centred learning environment, which emphasizes collaboration, personalization and feedback, will be fulfilled.

Strong support for a learner-centred E-Learning environment is essential in corporate organizations and the higher education sector where E-Learning involves knowledge and skills acquisition. It is in these two sectors that E-Learning is believed to have considerable potential in developing countries. This chapter discusses the basic concept of E-Learning, its requirements and advantages in the scope of developing countries.

1. Introduction

The telecommunications and knowledge revolution enabled greater and faster human communication and collaboration and led to fundamentally new forms of economic activity that "produced the Knowledge Economy and required basic changes in education" (Harasim, 2000). Regarding developing countries, what we need to do in terms of education is to move forward in line with globalization and the knowledge economy.

One of the key problems that developing countries face relates to widespread illiteracy. An estimated 847 million people are illiterate, including 20–40% of the populations of Africa, Asia, the Arab States and North Africa (UNESCO, 2002). In developing countries with large populations, traditional modes of education lack sufficient resources to ensure that all have access. Alternative solutions for education, such as decentralized units of educational institutions, must be explored (Khan *et al.*, 2001). However, this involves additional costs and may not be applicable to some developing countries. A solution to this problem is to utilize distance teaching or education wherever the decentralized support in terms of teaching and economic resources is insufficient (Khan *et al.*, 2001).

Web-based instruction today is based on behaviourism, viewing the learner as an empty vessel waiting to be filled. Distance learning educators should acknowledge constructivism as the new paradigm for learning and must also be willing to shift their teaching practices for consistency and constructivist learning (Morphew, 2002).

This chapter:

(i) provides an overview of the E-Learning domain, examining its origins from two perspectives—first, E-Commerce related issues and second, the use of technology in distance education.

(ii) investigates E-Learning environments and related current research from supporting tools to pedagogical aspects of learning, grounded on student-centred learning theories. The main focus is on the integration of the constructivist learning theory and use of technology in education, and more specifically on the identification of those elements important for establishing an E-Learning framework.

(iii) provides implementation examples of E-Learning in developing countries.

(iv) discusses tools used for E-Learning support and presents a brief critique of current systems.

2. From E-Commerce to E-Learning

This section gives an overview of the origins of E-Learning from two perspectives. From a business-oriented point of view, E-Learning is seen as part of the Business-to-Customer E-Commerce transaction model. Secondly, E-Learning is considered from the perspective of using technology in distance education.

2.1 E-Commerce and E-Learning

According to Schneider and Perry (2001), there are several transaction models for E-Commerce supporting transactions and business processes, such as Business-to-Business (B2B), Business-to-Customer (B2C), Consumer-to-Consumer (C2C), Business-to-Employee (B2E), Government-to-Business (G2B) and Government-to-Citizens (G2C). Electronic Data Interchange, Extranet and Electronic Bidding are examples of applications for B2B, while Cyber banking, Travel and Tourism, and Online Publishing are examples of B2C. The focal point for differentiating between B2B or B2C is by identifying the entities involved in the transactions. B2B implies both the sellers and buyers are business corporations, while B2C implies that the buyers are individual consumers (Turban *et al.*, 2003).

B2C transactions can be divided into two—transactions which involve products or services. Products include transactions such as online shopping while services include broker-based services, electronic job recruitment and online publishing and knowledge dissemination. Virtual Teaching is an example where knowledge dissemination is applied through electronic publishing (Turban *et al.*, 2003). Virtual Teaching encounters the use of computer-mediated communication technology and encompasses E-Learning, distance education and open education. Figure 4.1 indicates the position of E-Learning in the B2C transaction model.

2.2 Use of Technology in Education and E-Learning

A number of artificial intelligence techniques have been identified for use in developing E-Learning systems, covering both delivering the learning content and providing online assessment. In this way, we can ensure that these systems incorporate different learning styles. Such techniques are the Intelligent Tutoring System (Brusilovsky, 1999), adaptive hypermedia (Brusilovsky and Maybury, 2002), the nearest neighbour algorithm (Shih and Lee, 2001), the Bayesian Network algorithm (Wolf, 2002), and case-based reasoning (Shih and Lee, 2001; Gilbert and Han, 1999).

Brusilovsky (1999) reviewed three types of adaptive and intelligent technologies for Web-based Adaptive and Intelligent Educational Systems: (i) Intelligent Tutoring System technologies, (ii) Adaptive hypermedia technologies and (iii) Web-inspired technologies. The case-based reasoning methodology is used in developing adaptive online learning systems (Shih *et al.*, 2001; Gilbert and Han, 1999). Gilbert and Han (1999) introduced an adaptive E-Learning system, which accommodates different learning styles with the interaction of N instructors to one student. Shih *et al.* (2001)

applied the nearest neighbour computing algorithm. The online assessment of the system gives feedback to students by providing them with the most appropriate materials based on the results of their online assessments. The Bayesian Network algorithm is applied by Wolf (2002) to predict and recommend the most likely preferred options of media representations in encountering further learning material.

Figure 4.1
Positioning E-Learning within the B2C Transaction Model

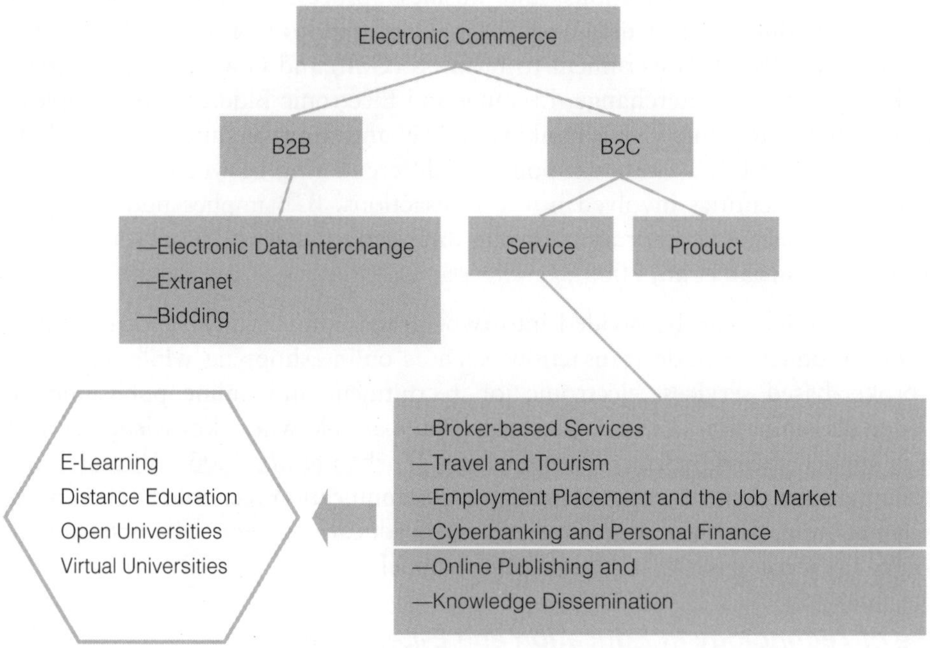

3. Distance and Open Education

3.1 Distance Education

Distance education (DE) can be defined as any type of educational situation in which the instructor and students are geographically separated by time and/or location. Although separated, there is a two-way interactive communication between the instructors and students in the process of acquisition of knowledge, skills and understanding. The two-way communication may be established using a wide variety of media including printed materials accessed through the postal service, telephone, television broadcast or computer network.

There are two conditions of distance learning, *asynchronous* and *synchronous*. Asynchronous instruction does not require the simultaneous participation of all students, thus enabling them to set their own schedules. Synchronous instruction requires real-time class participation through interactive TV sessions, video-conferencing or over the Internet.

3.2 Open Education

Learning is not limited only within higher education but will be a continuous activity. Learners in the 21st century will be life-long learners (Ben-Jacob *et al.*, 2000). One of the related learning theories which promotes life-long learning is the Open Learning theory, defined by Paine (1988, p. 59) as:

> "Both a process which focuses on access to educational opportunities and a philosophy which makes learning more client and learner-centred. It is the learning which allows the learner to choose *how to learn, when to learn, where to learn, and what to learn as far as possible* within the resource constraints of any education and training provision."

Part of this definition, where learners have the opportunity to choose "*when and where to learn*", shows that distance education is a subset of open learning. Open learning is an excellent approach when dealing with distant learners and is also highly appropriate in the case of online learners (Rosbottom *et al.*, 2000).

3.3 Use of Technology in Distance Education

Taylor (2001) suggests that there are five generations of distance education where each generation is associated with a specific model. The first generation is based on the Correspondence Model, which uses print technology. The second generation is referred to as the Multimedia Model, based on print, audio and video technologies. This model has proved very effective and continues in some Open Universities (Rosbottom *et al.*, 2000). Third generation DE, referred to as the Telelearning Model, is based on applications of telecommunications technologies to provide opportunities for synchronous communication. This generation is better known as computer-based distance learning (Rosbottom *et al.*, 2000). Use of computers in education in this generation initially involved stand-alone computer software known as Computer Aided Learning software; it then moved forward to the use of hypertext and multimedia techniques to enhance the capabilities of third generation systems.

Flexible Learning is the model for fourth generation DE, based on online delivery via the Internet. Finally, fifth generation DE, derived from the fourth, is referred to as the Intelligent Flexible Learning Model. Fifth generation DE emerged based on further exploitation of new technologies (Taylor, 2001).

Use of computers in DE started in the third generation. From this generation to the present, the use of technology is aimed at increasing the means of communication and interactivity. Interactive web-based instruction solved the problem of communication, where two-way communication is possible through communication tools such as news-groups, net-meetings and e-mails (Khan *et al.*, 2001). E-Learning emerged with network computer-based education. As Harasim has pointed out, online education has a long history, beginning with the invention of e-mail, and its development is intertwined with the history of computer networking (Harasim, 2000). In terms of the use of technology in DE, E-Learning originated through the use of networks in education, i.e., the Internet and specifically the World Wide Web (web). The web, which was invented in 1992, made online education increasingly accessible and allowed new pedagogical models to emerge. The web expands the range of disciplines that can be offered online through its ease of use and its capability of presenting multimedia (Harasim, 2000).

From this section, we note first that E-Learning can refer to web-based distance learning where technology is integrated with the concepts of both distance and open education. Shih (2002) states that distance learning is one of E-Learning's "special editions". Secondly, E-Learning can be a solution to the problems mentioned in the previous section. In the next section, we discuss current research on E-Learning, based on the constructivist learning paradigm.

4. E-Learning Environments

Dringus and Terrell (1999, p. 58) defined the online learning environment (OLE) as:

> "a distinct, pedagogically meaningful and comprehensive (space) by which learners and faculty can participate in the learning and instructional process at any time and any place. OLEs manifest a variety of technical *tools that support instructional delivery and communication in online formats*. In addition, dynamic delivery structures are embedded to enhance the instructional, learning and communication processes taking place."

Based on this definition, the aim of this section is to discuss both pedagogical and technical aspects of the E-Learning environment with respect to current research on E-Learning. Pedagogical aspects refer to the educational theory underlying the E-Learning environment, while technical aspects refer to the elements which need to be provided to students. Five main elements of the E-Learning environment have been identified in earlier research where no learning theory or paradigm was being considered (Iahad *et al.*, 2004)—Course Content, Assessment, Feedback, Course Management and Communication. These elements are described further below.

4.1 Course Content

This refers to the content provided by the instructor. In the case of open learning, the content is not necessarily provided by a single instructor, but may be outsourced from more than one experts in the area of study. Multimedia technologies are used in presenting the content to create an interactive environment.

4.2 Assessment

Assessment is needed in order to evaluate students on their understanding of the knowledge gained. Traditional methods of assessment such as quizzes, tests and examinations are assigned online. Online assessments are a means of evaluating students' knowledge in order to better accomplish the knowledge skills and concepts (Gusev and Armenski, 2001). Online assessment usually consists of multiple choices, true or false, fill in the blanks and essay type questions. Except for essays, automatic grading will be provided as soon as students submit the assessments online.

4.3 Feedback

Feedback is part and parcel of student-centred assessment. Student-centred assessment *promotes learning* instead of *monitors learning*. Huba and Freed (2000) observe that assessment that promotes learning must reveal to students an understanding of how their work compares to a standard, the consequences of remaining at their current level of skill or knowledge, as well as information about how to improve, if improvement is needed.

4.4 Course Management

This element includes facilities for both the instructor and students. It is common in E-Learning environments for tools known as Virtual Learning Environments (VLEs) such as WebCT and Blackboard to be used for providing a number of facilities. By using a course management tool in a VLE, instructors are able to achieve certain tasks such as granting student access for a specific module, input students course marks and post announcements. On the other hand, students are able to obtain their grades through personalized grade books, get informed about deadlines and view announcements regarding particular events, posted by the instructors

4.5 Communication

In an E-Learning environment, besides the interaction between the instructor and students, students should communicate among themselves. This is especially important in an open learning system where sharing of information among the students as well as with instructors is essential. In fact, in the student-centred paradigm, both the instructor and the student are involved in the learning process.

Of the five elements mentioned above, *communication*, which enables collaborative work, appears to be the most important element in an open E-Learning environment. Harasim stated that the principle of collaborative learning may be the single most important concept for online learning, since it addresses the strong socio-effective and cognitive power of learning on the Web (Harasim, 2000). This idea is also supported by Feldmann (2001), who claims that communication is an essential factor for a successful E-Learning environment. Collaboration means working together to accomplish shared goals where individuals seek outcomes beneficial to themselves and to other members of the group (Harasim, 2000). E-mail or Mailing list, Newsgroup, Audio and Video conferencing are some of the possible means of communication.

In the context of E-Learning, two major developments, constructivist theory and Internet technology, provided issues of concern in higher education (Tait, 1997). According to Hobbs (2002), the constructivist design process should be of paramount importance and should be analyzed by course designers prior to building a course. The following section discusses the paradigm shift from instructor-centred to student-centred learning, followed by current E-Learning research based on the constructivist learning theory.

5. From Instructor-Centred to Student-Centred

Traditional classrooms applied the objectivist or instructionist learning paradigm, based on Skinner's theory of transfer of knowledge from the instructor to the learners (Deubel, 2003). In contrast, the constructivist learning paradigm views learning as an active process rather than the result of a transmission of knowledge from programme to student (Herrington and Standen, 1999). In other words, for the constructivist model, learning is achieved by active construction of knowledge supported by various perspectives within meaningful contexts (Oliver, 2000).

In the instructionist learning approach, the importance of learning objectives is stressed. Objectives are sequenced into learning hierarchies, progressing from lower to higher order teaching (Ibrahim *et al.*, 2001). Huba and Freed (2000) described this paradigm of learning as instructor-centred. Emphasis is given to the acquisition of knowledge, which is frequently the memorization of information outside the context in which it will be used.

According to the constructivist learning paradigm, learners construct their own knowledge by actively participating in the learning process. Constructivist instructional developers value collaboration, learner autonomy, generativity, reflectivity and active engagement (Moallem, 2001). Following this, constructivist approaches employ different learning techniques such as active learning (Tait, 1997; Ibrahim *et al.*, 2001), co-operative learning, collaborative learning (Ibrahim *et al.*, 2001) and situated learning (Oliver and Omari, 2001).

6. The Constructivist E-Learning Paradigm

The focus of most E-Learning research is towards accommodating the constructivist theory as the learning pedagogy, where interaction and communication are supported through the use of computer-mediated-communication tools. In line with the emergence of new learning technologies, the theories of learning that hold the greatest sway today are those based on constructivist principles. Five examples of E-Learning systems based on the constructivist theory are presented as follows.

6.1 The Generic Model for Online Learning

Rosbottom *et al.* (2000) defined three learning environment elements in their generic model for online learning based on the Open Learning theory—*learning resources, management tools and virtual community.* Learning resources include the learning content which may be in the form of hypertext, graphics or multimedia, activities for students, formative assessments and downloadable materials which are the tools to accommodate learning by problem solving. They used the virtual learning environment software, WebCT, as the management tool which provides an explicit learning structure as well as a report generator. The virtual community which can be achieved by using both web-based resources such as notice boards, e-mail and online chat; and non-web-based resources such as telephone access and face-to-face access may help to alleviate the problem of high drop-out rates. All the elements are interrelated to each other through students' activities.

6.2 The Constructivist Information Model

Tait's Internet-based learning constructivist information model is itself based on three instructional design models—the cognitive apprenticeship approach, cognitive flexibility theory and anchored instruction (Tait, 1997). The model focuses on *information* and *support* provided by the instructor. Information is presented in an object-oriented hierarchy, which forms a study map. Using the map, students may start learning from any object and proceed to any connected objects. The study map includes links to outline syllabi for each topic and URLs for other recommended resources. Assessment involves authentic activities. Support from the instructor is seen as essential and complements the Internet-based learning.

6.3 The Constructivist Web Based Learning Environment

Oliver and Omari's (2001) research is based on problem-based learning and situated learning which emphasize authentic activities. All of these are grounded in constructivist theory. The tools they provide in their web-based learning environment are (i) *Course content,* broken into weekly topics, (ii) *Assessment* through problem-based group projects which involves authentic tasks, (iii) *Communication* or *interaction* through the use of bulletin boards and (iv) *Support* from tutors who have access to the problem solution and provide feedback.

6.4 The C3 Management System

Schneider *et al.* (2002) are convinced by the effectiveness of socio-constructivist pedagogies in education and struck by the apparent lack of widely deployed supporting tools. They include three elements, *Community, Content* and *Collaboration* in their C3 Management System. Their aim is to provide affordable support for innovative *scenarios,* which are a series of phases or activities. Students' activities include the tasks that they need to do and the roles that they play in a project-based learning model, which emphasizes student-centred instruction. The elements of Community and Collaboration are realized through these activities. For Content, students are required to seek information from various resources and disciplines in order to solve the assigned problems.

6.5 Design Guidelines for Material Supporting Constructivism

Oliver (2000) explores some of the issues surrounding the design and development of Web-based materials and provides a guideline for designing online learning materials which support constructivism. The guidelines are (i) Choose meaningful contexts for the learning, (ii) Choose learning activities ahead of the content, (iii) Choose open-ended and well structured tasks, (iv) Make the resources plentiful, (v) Provide support for the learning and (vi) Use authentic assessment activities. The first three elements are centred on authentic learning contexts and activities, which reflect real-life settings. The fourth element is concerned with providing students with extensive resources in order for them to be able to deal with the authentic activities. Oliver suggested two types of support, which are peer support obtained from collaborative and co-operative learning through the use of tools such as e-mail, bulletin boards and chat rooms. The second type of support is structured scaffold and support provided by the designer, such as by including FAQ sites, examples of other students' works and links to external sites. Providing authentic assessment refers to assessment which is closely tied to the selected activities; students must take particular care with learning activities that are in some way related to the planned assessment (Oliver, 2000).

7. An E-Learning Framework

In this section, a comparative analysis is made of the elements included in each of the E-Learning approaches discussed in the previous section.

7.1 Comparative Analysis and Findings

Table 4.1 shows that all approaches included *Content* their E-Learning environment, presented to the students from multiple resources. Four approaches included *Assessment,* in the form of authentic assessment; most involved problem-based or project-based assignments. Three approaches included *Communication* as one of the explicit elements, while two included *Community. Activities* were given the most emphasis and were included by all the researchers. Next, the element *Management* is included by two researchers. Finally, four researchers included *Support* as one of the elements for the E-Learning environment.

The conclusion is that learning should be centred on students' activities where students use the content or resources and communicate and collaborate in order to solve real-life problems assigned to them through assessments. Interaction is achieved through communication and collaboration and then forms learning communities.

Table 4.1
Comparative Analysis of E-Learning Research
Accommodating the Constructivist Model

	Content	Assessment	Activities	Communication	Community	Management	Support
Rosbottom *et al.*, 2000	•	•	•		•	•	•
Tait, 1997	•	•	•				•
Oliver and Omari, 2001	•	•	•	•			•
Schneider *et al.*, 2002	•		•	•	•	•	
Oliver, 2000	•	•	•	•			•

Support is one of the elements which was given most emphasis. Support, which can be either from the instructors or the institution, can be viewed as a type of interaction. One reason for providing support to students is to ensure that the feeling of isolation is minimized. Studies show that interactivity is critical to successful distance education systems (Starr, 1998).

7.2 E-Learning Frameworks

Dringus and Terrell's (1999) framework for DIRECTED online learning environments includes eight key elements—*Delivery, Interaction, Resources, Evaluation, Culture, Technology, Education* and *Design*. The elements do not stand on their own but need to be integrated in order to support an effective online learner-centred paradigm.

In other research, Khan (2001) included eight factors in his Framework for Web-Based Learning which he viewed as the factors that can encompass various online learning issues and provide guidance in the design, development, delivery and evaluation of flexible, open and distance learning environments. The factors are—*pedagogical, technological, interface design, evaluation, management, resource support, ethical* and *institutional*.

7.3 Introducing the SDE-CAF E-Learning Framework

A generic E-Learning framework from the viewpoints of both students and instructors is proposed (see Figure 4.2). The framework is named SDE-CAF after the six key concepts of this research, namely—Strategy (S), Development (D), Evaluation (E),

Content management (C), Assessment (A) and Feedback (F). It is based on the initial results from the comparative analysis of five pieces of research into constructivist E-Learning, and on the study of both previous E-Learning frameworks, where five elements were identified for an E-Learning environment: Content, Assessment, Feedback, Management and Communication.

Figure 4.2
The SDE-CAF E-Learning Framework

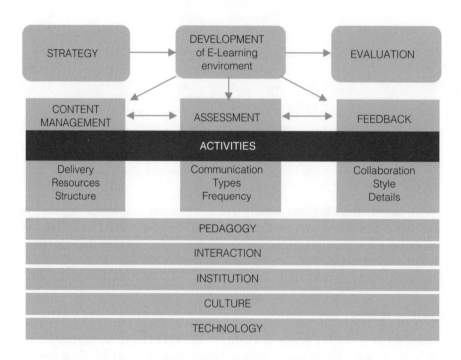

The three main factors essential to the whole process of E-Learning are *Strategy*, the *Development* of the E-Learning environment itself and *Evaluation* (SDE). In terms of systems development, Strategy reflects the activities of capturing the requirements in creating an E-Learning environment, Development includes the process of designing and implementing the functionality provided by the system and Evaluation involves the process and activities of testing the environment while evaluating its technological and pedagogical effects. The main factor which will be discussed here is the development of the E-Learning environment.

The three main elements for the E-Learning environment have been identified as *Content Management*, *Assessment* and *Feedback* (CAF). Contrary to some approaches (Dringus and Terrell, 2001), feedback is here considered as a distinct element of an E-Learning environment which must stand on its own. The three elements are centred on a number of defined *activities*.

Content Management is concerned mainly with the decisions made by instructors in delivering and structuring content as well as providing extra student resources. Assessment in a student centred E-Learning environment can be an integration of two types of assessment namely quantitative and qualitative (Kendle and Northcote, 2000). Quantitative assessment usually has the form of Multiple Choice Questions while qualitative assessment involves collaborative course assignments usually requiring the use of computer mediated communication tools such as Bulletin boards. Instructors also need to consider the frequency of assessments provided. Contrary to some approaches (Dringus and Terrell, 2001), feedback is considered here as a distinct element of an E-Learning environment. Different styles, timely and detailed feedback provided to students is essential in delivering quantitative type of assessment within a student centred E-Learning environment (Iahad *et al.*, 2004). Through collaboration, students also receive feedback and support from peers and the instructor.

Aside from these elements, five *concepts* which influence the whole E-Learning environment have been identified which are *Pedagogy*, *Interaction*, *Culture*, *Technology* and *Institution*.

8. E-Learning in Developing Countries—Examples of Good Practice

8.1 University of Pretoria, South Africa *(www.up.ac.za/telematic)*

The Department of Telematic Learning and Education Innovation is in charge of open and distance learning programmes at the University of Pretoria. Information from the university's website shows that it has 30,000 students registered as distance learners with about 25,000 studying for a further diploma in education management. About 3,500 postgraduate distance learners receive study material electronically, via e-mail, websites and television.

The department uses the technology-enhanced flexible learning or telematic education paradigm. This paradigm integrates three delivery modes—contact tuition, paper-based distance education and electronic education. This is in line with the strategic foci of the department, education innovation and electronic education.

Delivery for the electronic education is in the form of interactive television and video conferencing, interactive multimedia programmes and the Virtual Campus Solution. Interactive Television and Video Conferencing are used for lecture transmission and class discussion where students will be able to communicate with the lecturers through the use of open telephone lines. The multimedia programme enables students to interact with the subject matter at their own pace. The Virtual Campus solution is software which provides (i) administrative services, (ii) Web-based courseware and instruction where tutoring, assignments, assessment and accreditation are included, (iii) academic services and (iv) communication facilities.

8.2 Regional IT Institute, Cairo *(www.riti.org)*

The Regional IT Institute (RITI), established in 1992, is a leading training institution in Cairo, Egypt in the area of Information and Communication Technology. RITI has delivered 870 executive training programmes; 480 postgraduate students are enrolled and 300 students have already graduated. Three types of programme are offered which are (i) Academic Degree Programmes (ADP), (ii) Executive Development Programme (EDP) and (iii) IT Certified Programmes. Courses are offered in the areas of business, management, and information and communication technology.

For the first type of programme, ADP, the courses offered are joint delivery with other universities, such as Middlesex University, UK and the University of Louisville, USA. As an example, one of the collaborative programmes with Middlesex University is the MSc in Electronic Commerce, delivered through Middlesex's Global Campus programme.

Student learning is supported by three elements—the Learning Support Centre, CD-ROMs and the Internet. The Learning Support Centre functions as a centre for tutorial sessions, maintains a library of technical journals relevant to the courses offered, as well as an access centre for students which can be contacted by e-mail, telephone or fax for any other support. The CD-ROMs contain the course content and course work. The Internet is mainly used for the students to access e-mails and bulletin boards which will be used extensively by the student for interaction with staff and peer-to-peer communication and collaboration.

8.3 Universiti Tun Abdul Razak, Malaysia *(www.unitar.edu.my)*

Universiti Tun Abdul Razak (UNITAR) is the first virtual university in Malaysia, established in 1997. Initially 162 students were registered with UNITAR in 1998; to date UNITAR has 7,500 registered students at seven study centres in Malaysia, and one in Cambodia. The virtual university offers 11 academic programmes in the field of Business Administration, Information Technology, Humanities and Social Science, from first degree to doctorate degree level.

UNITAR utilizes their E-Learning Academic Model, which comprises resource-based learning, and the collaborative learning model. The E-Learning Academic Model includes six main components: (i) Courseware, which can be in the form of CD-ROMs or on the Web, (ii) Course Management through its Virtual Online Instructional Online System (VOISS), (iii) Study Centres which provide support to students in the form of face-to-face tutorial meetings, (iv) Virtual Library and (v) the Customer Relationship Management (CRM) call centre which is meant to support students facing academic, technical and personal problems.

The Courseware and Virtual Library are elements of the resource-based learning model. On the other hand, the VOISS, Study Centres and CRM call centres are elements of the collaborative learning model. The courseware and VOISS are the main elements which facilitate student learning. The courseware provides students with the learning resources such as course content, past years' examination questions and non-assessed self-tests. Course management facilities including e-mail, discussion boards, assignments, quizzes and announcements, provided by VOISS.

9. Identifying Supporting Tools for E-Learning

The tools investigated here may fall under any of the following categories of course delivery tools:

- Courseware Management Software—Software which offers functionalities ranging from students' course registration or deletion, delivery of learning materials and assessments, generation of report for course instructors and tools for communication.
- Courseware Delivery Software—Functionalities provided by this kind of software are more focused towards the delivery of learning materials and assessment.
- Assessment Question Builder—This kind of software provides only the functionality of building assessment questions.

Certain steps were followed in the process of identifying the features provided by each product. An initial overview of each product's features was followed by an investigation of the corresponding websites. This second step involved (i) listing the features of products through their general description from the website, (ii) getting information provided from documentation such as press releases and (iii) by evaluating in certain cases trial versions of products or online demonstrations. In some cases, instead of identifying the features through online trial or demonstration of the products from their own corporate website, product trial or demonstration had been achieved through institutions using the products.

9.1 E-Learning Tools—An Overview

The *WebCT* (www.webct.com) Campus Edition is an example of a Courseware Management System. It supports course development (e.g., syllabus tool), course delivery (e.g., communication and collaboration tool, and assessment tool) and course administration (e.g., assignment submission and grade entry).

The *Blackboard Education Suite* (www.blackboard.com) consists of Blackboard 5 which is a Courseware Management System, Blackboard Community Portal and Blackboard Transaction. Blackboard 5 offers course and content authoring tools, a robust assessment engine, as well as synchronous and asynchronous collaboration tools.

The *Ucompass* (www.ucompass.com) E-Learning product is the Educator software which has rich features of content delivery and publishing, communication and evaluation software. Ucompass provides the online demonstration of Educator from three different perspectives—Teacher, Student and Administrator.

The *ToolBook Assistant* and *ToolBook Instructor* (home.click2learn.com) are two authoring products of click2Learn.com. ToolBook Assistant is suitable for instructors who are new to developing online courses, while ToolBook Instructor is more suitable for instructors who have programming skills.

The *eCompanion* (www.ecollege.com) is an application provided by eCollege in addition to eToolKit and eCourse. Features offered by eCompanion include grade book, document sharing, calendar, and syllabus.

Total Knowledge Management (TKM) (www.gen21.com) Systems is Generation 21's E-Learning solution. This product is quite different from other products as it introduces Dynamic Learning Objects. Some features of TKM are scheduling and registration, grade book maintenance, chat rooms for live discussions and hypertext links to a wide range of learning resources.

TopClass (www.wbtsystems.com), a product of WBT systems, is a Content Delivery and Management System, which offers the main facilities in supporting E-Learning. TopClass provides personalized learning, assessment and testing, and built-in asynchronous collaboration.

ClassAct! (www.darasoft.net), is Web-based course management software from darasoft.net. The functionalities provided by this software are course materials, assessments, threaded discussion and assessments with online results posting and automated marking.

Interwise (www.interwise.com) offers E-Learning solutions through its product *iClass*. iClass provides facilities for communication, collaboration and learning through integrated video and audio, application sharing, surveys or tests.

Web Course in a Box is developed by MadDuck Technologies (www.madduck.com). Features offered by WCB are class information, announcements, schedules, students, learning links and utilities.

Oncourse (oncourse.indiana.edu) is a Web-based environment developed at Indiana University for teaching and learning. It provides an E-Learning environment with facilities such as course materials, chat engine, mail, online testing and conferencing.

Perception (www.questionmark.com/uk/home.htm), a product of Questionmark.com, is an example of an E-Learning tool which provides the function of developing assessments including tests, quizzes and surveys. Its basic features are question banks, online viewing of results, reports and item analysis; storage of answers, scores, and results in MS Access or MS SQL and the ability to give feedback.

QuestWriter (www.peak.org/qw) is a product of PEAK. One of its functions is to build quizzes, tests and other question-based activities which can be scored and recorded automatically to a grade book. Besides this it supports discussion groups and reading activities for monitoring students' progress through the reading materials.

Mallard (www.cen.uiuc.edu/Mallard) is an instructional environment developed with the financial support from the Department of Electrical and Computer Engineering at the University of Illinois at Urbana-Champaign. Mallard provides the course instructor with functions which enable organizing of the online course materials and testing students through interactive quizzes. The interactive quizzes provide instantaneous problem correction and grading.

WebTycho (tychousa1.umuc.edu) is a customized programme developed by the University of Maryland University College to facilitate course delivery via the Web. Some of the basic features of WebTycho are class announcements, syllabus, course content, conferences and assignment folders.

Based on this investigation of E-Learning tools, two categories of tools had been identified, *Learning Tools* and *Course Delivery Tools*. Learning Tools are tools which facilitate students in their process of learning. On the other hand, Course Delivery Tools refer to tools that support delivery of knowledge and evaluation of students' performance. Course Schedule, Announcement, E-mail, Threaded discussion, Chat, Glossary, Search, FAQ, Bookmark and Whiteboard are grouped into the Learning Tools while Course Syllabus, Course materials, Assessment, Online grading, Grade book and Resource links are grouped into the Course Delivery Tools. The next section gives a comparative analysis of the E-Learning tools discussed above, based on the two categories of tools; see Table 4.2.

9.2 Critique of E-Learning Tools

Almost all E-Learning tools provide the *Course Delivery Tools*; 14 of them included Syllabus, Learning materials and Assessment; and 13 included Online grading, Grade book and Resource links. This shows that the Course Delivery Tools play an important role in delivering a course online.

For the *Learning Tools*, E-mail and Threaded Discussion are the features included by most of the E-Learning tools (12), followed by the Announcement (11) Chat Engine (9), Bookmark (8), Schedule (7), FAQ and Search (6), Whiteboard (5) and finally Glossary (2).

Based on this analysis, all the features in the Course Delivery Tools need to be included in the online course. For the features in the Learning Tools, E-mail, Threaded Discussion, Chat Engine and Whiteboard perform as a medium for communication, which enables collaborative work. These features can be included in the system to support the student-centred paradigm. However, priority can be given to features that gained the highest score, which are the E-mail and Threaded Discussion.

Eleven of the tools analyzed included "Announcement", shows that it is an essential feature. The feature Bookmark was included by more than half (8) of the tools analyzed, so it can be concluded that it is preferable to have this feature. Although less than half of the E-Learning tools included Schedule, this feature must be included as it is important to ensure that students are "on track" in the learning process. This also applies to the FAQ feature. Besides providing the students with information on the learning materials, just like a HELP file, the FAQ guides the students so that they are not "lost" while exploring the online course.

Table 4.2
Comparing Features of Course Development Tools

Feature\Tool	1	2	3	4	5	6	7	8	9	10	11	12	13	14	15
Learning Tools															
Schedule	●		●	●			●					●	●		●
Announcement	●	●	●		●		●	●			●	●	●	●	●
E-mail	●	●	●		●	●	●	●	●	●	●	●	●		
Threaded discussion	●	●	●		●	●	●	●	●		●	●	●	●	
Chat	●	●	●			●	●	●		●			●	●	
Glossary	●		●												
Search engine	●	●	●				●	●		●					
FAQ	●	●	●				●					●			●
Bookmark	●	●	●	●	●	●		●			●				
Whiteboard	●		●		●	●				●					
Course Delivery Tools															
Syllabus	●	●	●		●	●	●	●	●	●	●	●	●	●	●
Learning materials	●	●	●		●	●	●	●	●	●	●	●	●	●	●
Assessments	●	●	●	●	●	●	●	●	●	●	●	●	●		●
Online grading	●	●	●	●	●	●	●	●	●	●	●		●		●
Gradebook	●	●	●	●	●		●	●	●	●	●		●	●	●
Resource links	●		●		●	●	●	●	●	●	●	●	●	●	●

1. Webct, 2. Blackboard, 3. Educator, 4. QuestionMark, 5. Questwriter, 6. ToolBook,
7. ECompanion, 8. Generation 21, 9. ClassAct!, 10. iClass, 11. TopClass,
12. Web Course in a Box, 13. Oncourse, 14. Mallard, 15. WebTycho

Finally, for the Search engine and Glossary, one of them may be included. However, the Search engine will be beneficial to the students as it provides them with a wider range of results compared to the fixed terms defined in the Glossary.

The findings of the above critique show that there is a plethora of available E-Learning features for instructors to use. The evolution of information and communication technologies could be the key element for achieving the transfer of skills and knowledge through the education systems of developing countries either by introducing them to E-Learning established networks with cores residing abroad or by encouraging the development of the necessary infrastructure and the exploitation of the existing paradigms and techniques.

10. Conclusions

This chapter focused on supporting E-Learning environments under the perspective of their potential applicability to developing countries. Certain important concepts were discussed in detail and more specifically the constructivist learning theory and the transition from an instructor-centred to a student-centred paradigm. An early contribution of this chapter was to clarify the positioning of E-Learning within the context of today's "e-everything" world from both a business and technological perspective.

By discussing the key issues of distance and open education, this chapter eventually established some guidelines of good practice for the development of an E-Learning environment by identifying the key elements of (i) course content, (ii) assessment, (iii) feedback, (iv) course management and (v) communication. Current work particularly focuses on the constructivist E-Learning paradigm where interaction and communication are supported through the use of computer-mediated-communication tools. Furthermore, emphasis is given to the transition to a more student-centred approach towards learning with instructors have the role of providing flexible content that can be easily personalized to individual needs.

The key contribution of this chapter is the introduction of the SDE-CAF E-Learning framework based on the findings of an overview of E-Learning methodologies and an investigation of existing proposed E-Learning frameworks. The suggested framework identifies three essential factors for the E-Learning process, namely *Strategy*, the *Development* of the E-Learning environment itself and *Evaluation* (SDE). It also identifies three main elements for the E-Learning environment development which are *Content Management*, *Assessment* and *Feedback* (CAF).

Three examples of success stories of E-Learning in developing countries are also briefly presented from South Africa, Egypt and Malaysia. Finally, the findings of a detailed investigation in E-Learning support tools are discussed, focusing on courseware management software, courseware delivery software and assessment question builders. Fifteen different tools are described and a comparative analysis of their features is included.

References

Ben-Jacob, M. G., Levin, D. S. and Ben-Jacob, T. K. (2000) The Learning Environment of the 21st Century, *International Journal of Educational Telecommunications*, **6**, 3, 201–211.

Brusilovsky, P. (1999) Adaptive and Intelligent Technologies for Web-based Education," *Intelligent Systems and Teleteaching*, **13**, 4, 19–25.

Brusilovsky, P. and Maybury, M. T. (2002) From Adaptive Hypermedia to Adaptive Web, in: Brusilovsky, P. and Maybury, M. T. (Eds.), *Communications of the ACM*, **45**, 5, 31–33.

Deubel, P. (2003). An Investigation of Behaviorist and Cognitive Approaches to Instructional Multimedia Design, *Journal of Educational Multimedia and Hypermedia*, **12**, 1, 63–90.

Dringus, L. P. and Terrell, S. (1999) The Framework for DIRECTED Online Learning Environments, *The Internet and Higher Education*, **2**, 1, 55–67.

Feldmann, B. (2001) Communication—The Essential Factors for a Successful E-Learning Environment, *Proceedings of SSGRR, Conference for Advances in Infrastructure for Electronic Business, Science and Education*, www.ssgrr.it/en/ssgrr2001/papers/Birgit%20Feldmann.pdf

Gilbert, J. E. and Han, C. Y. (1999) Adapting Instruction in Search of 'a Significant Difference', *Journal of Network and Computer Applications*, **22**, 3, 149–160.

Gusev, M. and Armenski, G. (2001) New Frontiers for Testing in E-Learning, *Proceedings of SSGRR, Conference for Advances in Infrastructure for Electronic Business, Science and Education*, www.ssgrr.it/en/ssgrr2001/papers/Marjan%20Gusev.pdf

Harasim, L. (2000) Shift Happens: Online Education as a New Paradigm in Learning, *The Internet and Higher Education* **3**, 1, 41–61.

Herrington, J. and Standen, P. (2000) Moving from an Instructivist to a Constructivist Multimedia Learning Environment, *International Journal of Educational Multimedia and Hypermedia*, **9**, 3, 195–205.

Hobbs, D.L. (2002) A Constructivist Approach to Web Course Design: A Review of the Literature, *International Journal on E-Learning*, **1**, 2, 60–65.

Huba, M. E. and Freed, J. E. (2000) *Learner-Centred Assessment on College Campuses*, Allyn and Bacon, USA.

Iahad, N., Dafoulas, G. A., Kalaitzakis, E. and Macaulay, L. A. (2004) Evaluation of Online Assessment: The Role of Feedback in Learner-Centered E-Learning, *Proceedings of the 37th Hawaii International Conference on System Sciences*, Kona, Hawaii, USA.

Ibrahim, D. Z., Silong, A. D. and Abu Samah, B. (2001) Practices That Facilitate Learner Control in an Online Environment, *Workshop on Developing Effective Online Delivery System for Institutions of Higher Learning*, Shah Alam, Malaysia.

Kendle, A. and Northcote, M. (2000) The Struggle for Balance in the Use of Quantitative and Qualitative Online Assessment Tasks, *17th Annual Conference of the Australasian Society for Computers in Learning in Tertiary Education (ASCILITE)*, Coffs Harbour, New South Wales, Southern Cross University Press, 1, 531–540

Khan, B. H. (2001) A Framework for Web-based Learning, in: Khan, B.H. (ed.) *Web-based Training*, Englewood Cliffs, NJ: Educational Technology Publications.

Khan, J. A., Khan, S. A. and Al-Abaji, R. H. (2001) Prospects of Distance Education in Developing Countries, *Proceedings of the International Conference on Millennium Dawn in Training and Continuing Education*, University of Bahrain, Bahrain.

Moallem, M. (2001) Applying Constructivist and Objectivist Learning Theories in the Design of A Web-Based Course: Implications for Practice, *Educational Technology and Society,* **4**, 3, 113–125.

Morphew, V.N. (2002) Web-Based Learning and Instruction: A Constructivist Approach, in: Lau, L. (ed.). *Web-Based Instructional Learning*, Idea Group, Inc., UK.

Oliver, R. (2000) When Teaching Meets Learning: Design Principles and Strategies for Web-based Learning Environments that Support Knowledge Construction, *Proceedings of the ASCILITE 2000 Conference*, www.ascilite.org.au/conferences/coff00/papers/ron_oliver_keynote.pdf

Oliver, R. and Omari, A. (2001) Exploring Student Responses to Collaborating and Learning in a Web-Based Environment, *Journal of Computer Assisted Learning*, **17**, 1, 34–47.

Paine, N. (1988) *Open Learning in Transition: An Agenda for Action*, Cambridge: National Open College.

Rosbottom, J., Crellin, J. and Fysh, D. (2000) A Generic Model for Online Learning, *Proceedings of the 5th Annual SIGSE/SIGCUE ITiSCE Conference on Innovation and Technology in Computer Science Education*, Helsinki, Finland, 108–111.

Schneider, G. P. and Perry, J. T. (2001) *Electronic Commerce*, Second Edition, Course Technology, Canada.

Schneider, D., Synteta, P. and Frété, C. (2002) Community, Content and Collaboration Management Systems: A New Chance for Socio-Constructivist Scenarios?, *Proceedings of the 3rd Congress on Information and Communication Technologies in Education*, Rhodes: Greece.

Shih, T. K. (2002) Distance Education Technologies: Current Trends and Software Systems, *Proceedings of the First IEEE International Symposium on Cyber Worlds*, 38–46.

Shih, B. Y. and Lee, W. I. (2001) The Application of Nearest Neighbor Algorithm on Creating an Adaptive On-line Learning System, *Proceedings of the 31st ASEE/IEEE Frontiers in Education Conference*, Reno, NV, NV., Vol 1, pp 10–13

Starr, D. A. (1998) Virtual Education: Current Practices and Future Directions, *The Internet and Higher Education*, **1**, 2, 157–165.

Tait, B. (1997) Constructive Internet Based Learning, *Active Learning*, **7**, 3–8.

Taylor, J. (2001) Fifth Generation Distance Education, *E-Journal of Instructional Science and Technology*, **4**, 1. www.usq.edu.au/electpub/ejist/docs/old/vol4no1/2001docs/pdf/Taylor.pdf

Turban, E., Lee, J., King, D. and Viehland, D. (2003) *Electronic Commerce 2004: A Managerial Perspective*, Prentice Hall.

UNESCO (2002) Regional Adult Illiteracy Rate And Population By Gender, www.uis.unesco.org/en/stats/statistics/literacy2000.htm

Wolf, C. (2002) iWeaver: Calling Towards an Interactive Web Based Adaptive Learning Environment to Address Individual Learning Styles, *Proceedings of the Interactive Computer Aided Learning Conference*, Villach, Austria, September 25–27.

About the Authors

Dr. Dafoulas is a senior lecturer in Business Information Systems, at the School of Computing Science, Middlesex University. He is also a visiting lecturer in Interactive Systems Design, at the Computation Department, University of Manchester Institute of Science and Technology (UMIST). His research interests lie where Information Systems meet Human Computer Interaction. More specifically he is concerned with the investigation of human aspects in information systems development (e.g., communities of practice), effects of computer supported co-operative work in software engineering (e.g., distributed team working), computer mediated communication (e.g., distance and E-Learning) and Information and Communication Technologies (e.g., Internet computing and E-Commerce). His teaching primarily focuses on Systems Analysis & Design, e-Business & Internet Commerce and Java Programming.

Dr. Dafoulas has authored several research papers for refereed journals and peer-reviewed international conferences, mainly in the fields of software engineering and computer-supported co-operative work. He has also acted as a member of program and organizing committees for a number of international conferences. The most recent projects that he has been involved with, were distributed team working in software engineering, allocation of e-contracts in virtual software teams, development and evaluation of online communities of practice and the computer-assisted assessment and communication patterns in E-Learning. Research partners include British Telecom, IBM, The Department of Trade Industry, and the Global Campus at Middlesex University.

Noorminshah Iahad had her undergraduate studies at the University Malaya, Malaysia and gained her BSc in Information Technology in 2000. She had been appointed as a tutor for two semesters at the Information Systems Department, Universiti Teknologi Malaysia, Malaysia in 2001. She then pursued her studies at the University of Manchester Institute of Science and Technology (UMIST), United Kingdom and gained an MSc in Information System Engineering in 2002. She is now doing her PhD at the same university. She gained experience in developing Web applications during her undergraduate studies and in E-Learning support environment research while doing her MSc dissertation. Her research interests are in Computer Supported Collaborative Learning, Online Communities, and Computer Aided Assessment.

TELECENTRES

Call them what you will; Multipurpose Community Telecentres, information kiosks, information shops, Internet centres; we will use the generic term "telecentres" in this section. But we are definitely not talking about cyber cafés. Whilst telecentres provide shared access to Information and Communication Technologies (ICTs); i.e., computers and related equipment, telephones, and the Internet, their principal purpose is to promote local development not to make a profit. Telecentres address the digital divide and the asymmetric distribution of global access to information and to ICTs. Whilst telecentres are commonly promoted as the only practical means of reducing such unevenness, their effectiveness is only realized within workable development strategies, as Balaji describes in the first chapter in this section by describing the development role that telecentres can play.

Whilst cyber cafés also diffuse access to ICTs, potentially to poor people, their commercial imperatives inevitably pull them towards urban areas of high population density, leaving the rural populations where the vast majority of the poor live unserved. Of course if a telecentre can generate income, it has a better chance at survival, but many question why the poor should be expected to pay for their own development and why the financial sustainability of development telecentres should be treated differently from other public services like, libraries, schools, universities, roads, and even rail networks and airlines that are unable to pay for themselves.

The results of the many experiments with telecentres are mixed; some have demonstrated considerable benefits for their target audiences; others are struggling with fragile connectivity and uncertain communities. Although very few have achieved self-financing sustainability, that is not to say that they have failed, or that they are unable of eventually generating income sufficient to cover their operating costs. The following figure depicts the common types of telecentres according to the two most-often referred to success criteria; self-financing sustainability and development outcomes.

Telecentres are commonly perceived and assessed along the two dimensions in the figure; those that seek to be self-financing and those that seek to induce development. Telecentre assessments frequently refer to only one of these dimensions; for example, assigning a label of failure to a telecentre that has been responsible for inducing a desirable development outcome, possible even saving a life, but that fails to generate sufficient income to cover its costs. Conversely, successful telecentres are sometimes depicted as those that are commercially viable as a stand-alone operation, but which deliver basic communication and entertainment services that are useful but which fail to have a detectable impact in terms of community development. In the second chapter, Harris synthesis some experiences from observations on what makes a telecentre successful.

Types of Telecentres and Success Criteria

High

Experimental
Research
Telecentre

Mature
Development
Telecentre

Development
Outcome

Subsidized
Access
Points

Cyber
Café

Low Self-financing Substainability High

A further consideration that the model in the figure brings out relates to the dynamics of telecentre development. Many of the telecentres that might be classified within the model are capable of moving from one category into another over time. Research telecentres are often established in order to demonstrate their social benefits. Self-financing sustainability is a secondary consideration. Additionally, subsidized installations are often established to be handed over to private, profit-making ownership, after an initial period of operation that serves to demonstrate the potential and to stimulate the demand for access to ICTs. In both cases the period of time over which such a change occurs depends on a variety of factors that are often beyond the control of the project's initial promoters.

Most implementations will begin life in one of the "low" quartiles of the model, so that specific mechanisms are required to ensure that they move into the "high" quartiles over time. A lack of understanding of the dynamics of the model in the figure often leads to premature and inappropriate classification of an installation as a success or a failure. Experience indicates that development oriented telecentres require at least three to four years before generating sufficient revenue to cover operational costs, and under the right conditions, and given enough time, development telecentres can do this. Progress to the mature telecentre requires a programme that adopts a balanced consideration of both the financial and development dimensions of the model, as well as patience with recipient communities who will set their own time-scales irrespective of the schedules of donor agencies and government departments.

Under the right conditions, a well-run community based telecentre with just one or two computers, and possibly a few other items of equipment, can generate significant benefits for entire communities; assisting farmers, teachers, health workers and local businessmen and traders. Development telecentres operationalize the well-understood relationship between information and development, empowering the poor to access abundant information resources within processes that are largely under their own control. In partnership with sympathetic institutions, government and non-government as well as those in the private sector, the result can be to mobilize the potential of the vast resources that poor people possess and to overcome the previously significant barriers that they faced to making good use of them. Colle and Roman give some examples of how such partnerships can work in the third chapter in this section. In our final chapter for this section, Cocchiglia exposes all these issues, along with others, in his snapshot of a telecentre-based development programme in one developing country, Azerbaijan.

5. Explaining the Success of Rural Asian Telecentres

Roger W. Harris

*Consultants in Information and Communication Technologies
for Rural Development in Asia, Hong Kong*

harris38@netvigator.com

rogharris.org

Abstract

The global digital divide threatens to deprive millions of people of the benefits of Information and Communication Technologies. Most of these people live in rural parts of developing countries and they are unlikely ever to own their own computers or telephones. However, international aid agencies, governments and non-government organizations are becoming increasingly enthusiastic about the potential for generating rural development from community based telecentres. This report describes five telecentre projects that are concerned with bringing about social and economic development in rural communities in Asia. A success model of telecentres is developed, which is then applied to the five Asian telecentre case studies. Results indicate that the characteristics of communities emerge as the most potent influence on the success of community telecentres, yet are probably the least manageable.

1. Introduction

The developing world's lack of access to ICTs, often characterized as the digital divide, originates in several factors including—reliance on personal computers (PCs) that are too expensive for most developing country citizens to purchase; paucity of telecommunications, especially in rural areas where most developing country inhabitants live; low levels of literacy and education; and dependence on western keyboards for non-western scripts. Telecentres are community-based resources that provide shared public access to Information and Communication Technologies (ICTs) for the purpose of community development. Rural telecentres are springing up throughout Asia and they represent attempts to remove these barriers to access to ICTs. If they are to deliver on their development promise, it is important to understand whether, how and to what extent telecentres are able to overcome both the structural and human inhibitors to access.

Telecentres come with a variety of names and functions. Sometimes they are known as multipurpose community telecentres, village information shops or information kiosks. The services they offer include one or more of—telephone, e-mail, facsimile, photocopying, web browsing, information retrieval assistance, general purpose computing and computer training. A defining characteristic distinguishing community telecentres from cyber cafés is that community telecentres target community development from the public access to ICTs and information that they provide. Cyber cafés are generally only concerned with profit, and they are therefore mostly found only in high population density urban areas, although this does not preclude them from achieving community development.

Governments and international donor agencies are becoming increasingly interested in community telecentres as a means of diffusing ICTs, delivering useful information and accelerating development in the developing countries. By studying telecentres, we can learn how interventions aimed at removing access barriers to information are able to generate development. As Asian governments take an interest in delivering ICT access to their previously under-served rural populations, we need to know whether it will be possible to overcome the access barriers of the digital divide and whether a telecentre success theory can be synthesized that will be capable of assisting future implementations. This chapter proposes a model of telecentre success based on previous literature and on five case studies in which the author participated, either as project leader or as project evaluator.

2. Telecentres and Development

Telecentres are different from cyber cafés in that their purpose is to achieve development, rather than profit, so the success criterion in a model of community telecentre success is community development. Community development is locally definable, and it occurs when a community recognizes that it has moved from one state to another state that it considers more desirable. The concept of community development is central to an integrated approach to development in which the workings of a local economy are inseparable from wider social, political and cultural processes. However, community development has a tendency to become largely cosmetic unless it involves the active participation of the community in the planning stages of projects. Furthermore, participation is itself problematic as it often masks the differences between people; local heterogeneity is dissolved into vague notions of community, potentially disregarding important crosscutting divisions of class, gender and age which may lead to substantial differences in local views and interests (Gardner and Lewis, 1996).

In assessing community development, Whyte (1999) suggests several questions that should be asked:

- Is the telecentre a positive force for community development?
- Does it benefit some people more than others?
- Does it act as a catalyst for other positive initiatives and innovations at the local level?
- Does it help people to help themselves?

She further suggests a number of actions, notably—obtaining good baseline data on the community; measuring succeeding changes carefully; demonstrating a strong association between the telecentre and the economic and social changes found; and applying the argument of reasonableness in judging the likely direction of causality.

Colle (2000) suggests other issues for inclusion in a general theory of telecentre success, some of which relate to what Markus and Soh (2001) describe as structural conditions, which impact how IT innovations are implemented within different countries or regions. Structural conditions differ from country to country, and even from location to location within a country, but they are not necessarily related to dimensions of national culture. Therefore, valid explanations of differences in IT implementation activity require a careful assessment of relevant structural factors. Colle (2000) emphasizes the need for participation by local communities in order to make telecentres understood, but he says the concept of participation in telecentre development is not absolutely clear. Additional issues relate to the planning of telecentre activities, their product mix, the localization of knowledge and information resources, the nature of their start-up, the extent to which telecentres are networked among themselves, and telecentre financing (Colle, 2000). Table 5.1 summarizes some previous findings relating to telecentres and their success indicators.

Table 5.1
Previous Findings Relevant to Telecentres

Author(s)	Findings
Kanfi and Tulus (1998)	Definition of a telecentre as "a location which facilitates and encourages the provision of a wide variety of public and private information-based goods and services, and which supports local economic or social development".
Fleury (1999)	Telecentres, telecottages, teleports, community networking public access sites... these are all variations on the theme of supporting community access to new technologies and the applications which support locally defined development objectives.
Hudson (2001)	One approach to identifying objectives is to ask the various stakeholders: "What would make this a successful (telecentre) project?". Their answers might be, for example, that the project would: "provide people in the community with access to ICTs, train community residents in the use of ICTs".
Fuchs (2002)	Once the telecentre is up and running, its staff must court the community at large and introduce its members to basic computer skills and identify ways in which they might benefit from the facilities and services.
Hudson (2001)	For telecentres to have an impact on development, at least the following is required—Community access: The equipment must be conveniently located. The telecentre must be open at hours when people want to use it.
Roman and Colle (2002)	Themes that may provide starting points for generating hypotheses regarding successful telecentres (include)—the importance of raising awareness about information and ICTs as a valuable resource for individuals, families, organizations and communities, focusing on information services rather than on computers and the Internet alone to build a local institution more fully woven into the fabric of the community.
Benjamin (2000)	Pilot telecentre projects aim at advancing "the access of disadvantaged communities ..to modern information and communication technologies and to apply them to their own development priorities".

3. The Case Studies

Five telecentre case studies were conducted between November 1999 and June 2001. Cases are drawn from an initiative by the International Research Development Centre (IDRC), the PANAsia Telecentre Learning and Evaluation Group (PANTLEG) consisting of five Asian telecentre projects that have received funding from the IDRC.

The objective was to join the projects together in a closer partnership for mutual benefit (PANTLEG, 1999b). The projects are as follows:

(i)   e-Bario, Malaysia.

(ii)   MS Swaminathan Research Foundation (MSSRF) Village Information Shops, India.

(iii)   Foundation of Occupational Development (FOOD) Chennai, India.

(iv)   Multipurpose Community Telecentres, The Philippines.

(v)   Internet Information Centres, Mongolia.

Case descriptions are drawn from evaluations that were conducted by PANTLEG members, i.e., the project leaders plus an additional member from each project. The current author was project leader for the e-Bario project and took part as a team member of the PANTLEG team conducting the learning evaluations of the projects in India and Mongolia. He conducted an evaluation of the Philippine project as a consultant to the IDRC. Data collection for the cases consisted of site visits, interviews with project staff, telecentre operators and users, and direct observations. Interviews were conducted in groups, with the assistance of translators where necessary.

3.1 E-Bario

This is a telecentre in a village called Bario in Sarawak, which is one of the two Malaysian states on the island of Borneo. The State is characterized by its diffused population spread across hilly and forested terrain with under-developed infrastructure. The telecentre serves as a community ICT resource (Songan *et al.*, 2000). The settlement of Bario, with a population of around 1,000 people, is inaccessible by road. The community is made up of people of the *Kelabit* ethnic group, one of Sarawak's smallest among its 26 or so identifiable ethnic minorities. Forested mountains surround the plain in which the residents cultivate wet rice. People in the older part of the settlement live in a *longhouse*, the traditional form of dwelling in Borneo. A telecentre was established in 2000 with four PCs and two printers and its own electricity supply. Telekom Malaysia has installed VSAT (Very Small Aperture Terminal) satellite equipment to connect the telecentre to the Internet (Harris *et al.*, 2001). The project was instigated in 1998 by the author, who made nine site visits between April 1998 and November 2000, averaging 6 days per visit.

3.2 MS Swaminathan Research Foundation (MSSRF) Village Information Shops

MSSRF has established a series of rural telecentres in Pondicherry, southern India, where about 20% of families survive on less than US$1 per day. The project started in 1998 with the following objectives:

- To set up village information shops that enable rural families to access ICTs.
- To train youth and women in rural areas in operating the information shops.
- To train youth to generate locally relevant information.

- To maintain and disseminate information on entitlements to rural families.

A value addition centre was set up as an information hub in one village with dial-up accounts to the Internet. Five additional village centres have been connected to the hub. Trained volunteers run each centre. Many locally useful databases have been installed, based on local sources and stored in the local Tamil language. The centres receive an average of 12 visitors per day, among whom asset-less, ultra-poor families are the major users. Benefits from the centres include price information for farmers and wave conditions for fishermen (MSSRF, 2000). The PANTLEG team visited the project in November 1999, visiting five villages over a period of a week.

3.3 Foundation of Occupational Development (FOOD) Chennai

This project, based in Chennai, south India, enables local development agencies to use ICTs, especially in remote areas. The project established a remote area electronic network using packet radio modems to support telecentres in 10 remote sites, one of which is a featured case here. The village, a remote tribal community, used the tele-centre to establish an herbal cultivation business, obtaining the information they needed from the Internet and using the telecentre to market their products. The community had previously been engaged in snake catching, but was encouraged to switch to a more sustainable occupation that did not risk wiping out the snakes that kept the rat population down (FOOD). The telecentre has accumulated a considerable library relating to herb cultivation, enhancing the traditional knowledge it possessed regarding herbal remedies for its livestock. The herb business has extended from local gathering of wild herbs to their cultivation. The project was visited for a day by the PANTLEG group in November 1999, and they interviewed a group of telecentre staff and community representatives.

3.4 Multipurpose Community Telecentres

This project operates in four farming and fishing villages in northern Mindanao in the southern Philippines. The purpose is to test a pilot information and communication system that will support rural communities in achieving sustainable development. The four villages provide premises for their multipurpose community telecentre (MCT), along with staff, utilities and other supplies. Partner information providers from government and other institutions deliver information to the MCTs. Each MCT is staffed by at least 10 trained volunteers. MCT services include word-processing, printing, training and information referral services. Each MCT is expected to serve a cluster of five or more other villages. The output of the project is intended to be a generic model of operation for MCTs to be deployed across the rest of the country. Other outputs include information services for community development and expanded local capacity for co-operative organization, health care, agriculture, education and rural enterprises. The author was commissioned by IDRC to conduct an evaluation of this project. It involved visits to four village MCTs, involving interviews with telecentre operators, volunteers, community representatives and community members who use the facilities, as well as interviews with key project personnel in Manila.

3.5 Internet Information Centres

The purpose of this project is to deliver Internet access to rural areas in Mongolia. Mongolia is a large country with few (2.4 million) inhabitants, of whom nearly 25% live in the capital city Ulaanbaatar. Most of the remainder are engaged in a nomadic life-style, herding livestock across the grasslands. The telecommunications infrastructure is under developed, especially in rural areas, and the telephone network is based on obsolete Russian technology. Telecentres have been established in two provinces. They are connected to the Internet via a VSAT satellite system at a speed of up to 64Kbps. The main sponsor is the Open Society Institute (OSI) in Mongolia, sponsored by the Soros Foundation, and Datacom Co. Ltd. of Mongolia, a local Internet Service Provider. The IDRC are also in support. The telecentres, known as Public Internet Centres (PICs), provide free Internet connections to secondary schools, local government offices and non-governmental organizations (NGOs). These services are stipulated by the project sponsors. The PICs serve their members, who can be NGOs, such as a local women's group. Whilst NGOs enjoy subsidized usage, business users are charged for Internet access. The PICs provide—Internet access, with 6 PCs, e-mail, fax, Web hosting and design, and local telephone service.

The price for these services varies according to the type of service and according to the user. Several problems have been encountered:

- The electricity in rural areas is not stable.
- The telephone connection is unreliable and its speed is low.
- Language is a problem as most documents on the Internet are in English
- Rural PC penetration and computer literacy is low, compared to the urban population.

The PANTLEG group performed an evaluation of this project in June 2000. Three sites were visited in Erdenet, Choybalsan and Ulaanbaatar. Visits occupied a day at each location and included group interviews with telecentre staff, user groups and community representatives.

4. Case Study Findings

Arising from the case studies, some common factors were observed as being instrumental in achieving each project's outcomes.

4.1 Structural Conditions

Developing country governments are being encouraged to formulate national strategies for rural development by narrowing knowledge gaps through technology acquisition and distribution, education and training and expansion of access to technologies by de-regulation and privatisation (World Bank, 1999). Government policies and political leadership will determine the success of such policies. The participation of the major international donor and aid agencies such as the United Nations Development Programme (UNDP) or the Canadian Government's International Research Development Centre (IDRC) can heavily influence telecentre pilot projects and hence nation-wide roll-out programmes.

4.2 Community Characteristics

Telecentres serve communities but observation suggests that a variety of social factors influence the extent to which communities engage with and ultimately embrace telecentres as community resources (PANTLEG, 1999a). Technology is quickly embedded within contextual social factors and the outcomes are a result of their interactions. Some of the social factors in communities that have been observed to define outcomes of telecentres are:

- Development aspirations
- Capacity for learning
- Capacity for change and adaptation
- Ability to organize themselves and their assets
- Unity in decision making and in taking action
- Participation in development activities
- Harmonious internal relationships
- Strong personalities that motivate others

4.3 Telecentre Characteristics

In fulfilling the development role of the telecentre, staff provide training and assistance to community users. The role of the *infomediary* has been identified as an individual working in a telecentre and drawn from the community, who is capable of using ICTs in order to respond to requests from the community for information for solving some problems that might yield to an Internet enquiry. A variety of managerial and financial aspects relating to telecentre operation also influence outcomes, such as the ability to engage with the local information needs of the community and to generate some cost recovering revenue.

4.4 Information Characteristics

Information systems should be useful and useable. According to Colle (2000), a telecentre designed to support community development should be aggressive and creative in localizing its knowledge and information resources. Given the local specificity of telecentres, personality variables can also be expected to influence telecentre outcomes. They include, demographics, the influence of champions and the strength of personalities of opinion leaders within communities.

Table 5.2 summarizes the case studies. For ease of reference, the table supplies a brief description of the distinguishing characteristics of each case followed by an assessment of the level of success achieved by the telecentres. Each case study is summarized against the common factors listed above.

5. The Telecentre Success Model

Assessing the case studies within the framework of proposed success indicators leads to a model of telecentre success as shown in Figure 5.1.

Figure 5.1
Model of Telecentre Success

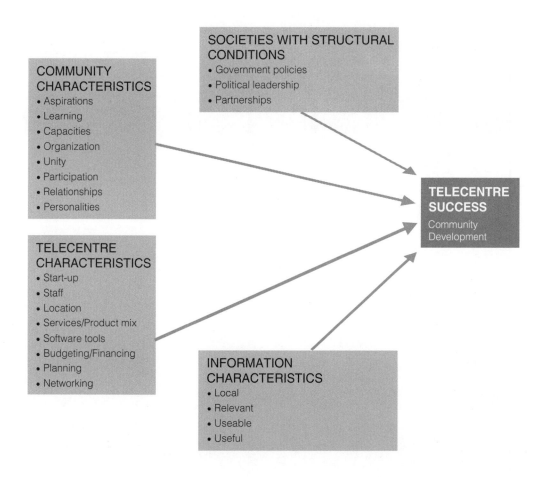

Table 5.2

Observed Case Study Outcomes

	e-Bario, Malaysia.	MS Swaminathan Research Foundation, (MSSRF) Village Information Shops, India.	Foundation of Occupational Development, (FOOD) Chennai, India.	Multipurpose Community Telecentres, Philippines.	Internet Information Centres, Mongolia.
Distinguishing Characteristics	A small, remote community. Traditional home of an ethnic group. Access is practical by air only.	Hub and spoke model of information delivery. All locations are rural.	One telecentre in a remote rural tribal community	Small rural coastal communities. Information from government agencies and other institutional support.	Mixed models of telecentre design supporting NGOs and providing public Internet access
Community Development at the Time of the Study	Early stages yet. Limited to e-mail, which is important.	Considerable benefits improving social well being	Re-focused commercial livelihood to sustainable herb cultivation	Developmental gains yet to emerge	Mostly improved communications
Structural Conditions					
Government policies	Supportive. Government aggressively pursuing ICTs.	Initially a hindrance, but lately coming round.	Neutral	Supportive and instrumental to project inception, but slow to assist.	Newly supportive
Political leadership	Substantive support for ICTs	Not a significant feature	Not a significant feature	Important for the project	Not a significant feature
Partnerships	Instrumental	With IDRC project partners, and others	Not important	Important	Significant
Community Characteristics					
Development aspirations	High	High	High	Low, as yet un-ignited	Low to medium, to be ignited
Capacity for learning	Willing and anxious to learn	Advanced	Capable. Focused on specific needs	Slow but willing	Willing but restricted to small groups
Capacities for change and adaptation	High, with diaspora support	Highly motivated	Proficient at what they need	Latent, potentially high	Not fully exploited
Ability to organize themselves	Loose but locally focused.	Considerable	Sufficient	Burgeoning	Proficient but with further potential

	e-Bario, Malaysia.	MS Swaminathan Research Foundation, (MSSRF) Village Information Shops, India.	Foundation of Occupational Development, (FOOD) Chennai, India.	Multipurpose Community Telecentres, Philippines.	Internet Information Centres, Mongolia.
Community Characteristics (continued)					
Unity in decision making and in taking action	Traditionally tight but emphasizes consensus finding and therefore tends to be slow to implement	Sufficient and maintained with help from the project staff. Significant hierarchy of influence in the communities.	Sufficient for tight focus on specific requirements	Considerable. Local officials well supported by the communities.	Diverse across various stakeholders
Participation in development activities	Community has equal rights with the project staff. Local steering committee makes decisions.	Shared responsibilities, but transiting towards the communities.	Self-governing	Community began as willing collaborators more than equal participators, but decisions are evolving toward to the communities	Communities are consultative and rather passive
Harmonious internal relationships	Close and deep based on common ethnic background and extensive extended family relationships	Complex, with a variety of social strata, including those based on caste.	Tight and ocused. Based on common ethnic affiliation, locally differentiated as "tribal" peoples.	Close and co-operative, based around community belonging	Varied and scattered throughout a range of stakeholders
Strong personalities that motivate others	Influential champions within the community have consistently supported the project, adding to its community acceptance.	Influence of young, capable volunteers is based on their achievements with the technology and the community contributions they make.	Individuals specialists with particular roles that support community use of the telecentre	Influential local leadership support plays a positive role in community acceptance	Local management and volunteers encourage usage
Telecentre Characteristics					
Started by community-based actors as opposed to outsiders	Research organization	Research organization	An NGO within the community	Government	NGO/Private organization
Staff pro-actively seek and disseminate useful information for the community	Local recruits, but also the diaspora for visioning applications and contributing to their development	Locally recruited and trained volunteers. Competent and confident with the technology. Women volunteers instrumental in bringing benefits to women members of the community.	Locally recruited volunteers with adequate skill to support the highly focused application.	Local school pupil volunteers, instrumental in spreading awareness but unable to address the use of the telecentre to solve community problems.	Local paid management and volunteer staff. Technically capable and able to spread awareness. More effort needed to transcend immediate benefits of email to target community development.

	e-Bario, Malaysia.	MS Swaminathan Research Foundation, (MSSRF) Village Information Shops, India	Foundation of Occupational Development, (FOOD) Chennai, India.	Multipurpose Community Telecentres, Philippines.	Internet Information Centres, Mongolia.
Telecentre Characteristics (continued)					
Central and easily accessible location	Central community lodging house, although temporary, is regularly used as a community meeting place.	Central locations, provided by the community. In one instance, part of the temple.	Central location, provided by the community, dedicated to the purpose.	Central location, provided by the community and associated with the existing administration.	Central, shared with government establishment, but not immediately accessible. On the first floor of a hotel in one instance.
Provides a mix of services and products that communicate useful information in a variety of ways	Development oriented communications and information provision planned in conjunction with the community	Development oriented, devised by the community and project staff. Mix of communications and information provision.	Communications and commercial information concerned with a single activity, herb cultivation, that the community specializes in.	Development oriented information provision, supplied by government and other agencies in a help desk arrangement that responds to community requests.	Mostly communications and web site development
Utilize software tools appropriate and adequate for the community	e-mail, Internet , word-processing.	E-mail, Internet. Spreadsheets, databases, word-processing, graphics, audio.	E-mail, Internet, databases.	E-mail, Internet, word-processing.	E-mail, Internet, word-processing.
Has sound budgeting and sustainable financing	Research funding and government grant	Research funding and government grant	Research funding and income generating	Research funding	Donor agency and income generation
Conduct adequate planning	Research based with community participation	Research based with community participation	Locally by a community based NGO	Research based with community co-operation	By the NGO/Private organization partnership
Actively network with other telecentres	Minimal as yet	Between themselves, within the five telecentres.	Focuses on their prime commercial activity	Within the project's four telecentres	Minimal as yet

	e-Bario, Malaysia.	MS Swaminathan Research Foundation, (MSSRF) Village Information Shops, India.	Foundation of Occupational Development, (FOOD) Chennai, India.	Multipurpose Community Telecentres, Philippines.	Internet Information Centres, Mongolia.
Information Characteristics					
Local; pertaining to matters of local interest	Not much yet, but there is a community-generated development agenda that focuses on local information.	Extensive local and regional information databases	Almost entirely local, relating to the cultivation and marketing of local herbal products.	Not much yet, information provision began much as a top-down programme, with some local consultation.	Little, but developing as local web sites get under way.
Relevant; pertaining to matters that are of interest to the community	Not yet, but expected.	Highly relevant to local needs and with varied topics	Highly relevant to local needs, but limited in range of topics.	Not much of local relevance	Not much of local relevance
Usable; information in a useable format	Not provided yet, but targeted to be so. Language will be an issue as there is no standard written form for the local language.	Distributed in local Tamil script, some available in sound files for those who cannot read. Also visual.	Local script and tightly focused on the needs of the main commercial activity	Some, but not much	Language is an issue, and there is not much available anyway.
Useful; information that the community can put to good use	Expected by the community to contribute to key areas of community concern: commerce, culture, education, and health.	Extremely useful, making significant contributions to community well-being in education, health, and commerce.	Extremely useful, making significant contributions to community well being in the form of commerce.	Not to a great extent yet	Not much yet

6. Discussion

In both the MSSRF and FOOD telecentre projects, the aspirations of the communities have been harnessed in a positive way, propelling them towards positive outcomes with their telecentres. Telecentre volunteer staff appeared motivated and proud in what they are able to achieve and the enthusiasm of the community was evident in the positive manner in which they described the operation of their telecentres. Such encounters also testify to the rate of learning that the communities have displayed through, for example, the ease with which the semi-literate volunteers are able to work with a western keyboard in entering Tamil script. Community capacities are clearly elevated and are sustained by the capability of the communities to organize and operate effective services from their telecentres.

The telecentres in the Philippines and Mongolia depict greater influences in the government policies compared to the other projects. The Philippine project is operated by a government agency and enjoys access to high government officers. The Mongolian project is closely allied to government officials at the local level, even providing subsidized services to them at one stage. It seems likely that this close affiliation will lead to additional support to compensate for external funding drying up.

The e-Bario case is strong on the community factors but logistics and power generation complicate technology management. Nevertheless, the community is motivated, united and highly resourceful (a pre-condition for survival in a remote locality) and there are grounds for optimism in supposing that the telecentre will make a substantial contribution to community well being.

Overall, the findings indicate some promise for the model of telecentre success to be able to account for the success of community telecentres. The factor about which the cases provide the least information relates to financial sustainability of telecentres. All of the telecentres in the cases began as experiments, supported in part by research funds, with financial sustainability as one of the research questions. None of the telecentres has yet achieved financial self-sustainability, although those in the FOOD case and in Mongolia are generating revenues. It is still too early to establish whether any of the telecentres in the cases will be capable of supporting themselves financially, or the extent to which some form of subsidy will be required in order to sustain them. The requirement for financial sustainability (not necessarily *self*-financing) is clearly critical for telecentre success, and whilst it is still unclear as to what conditions are required for telecentres to induce community development in a financially sustainable manner, it is clear that subsidies are required in most cases.

The apparent influence of community characteristics in telecentre success presents a challenge for ICT oriented development practitioners. In all the cases here, introducing the technology was not straightforward; remoteness and under developed infrastructures hindered its deployment. Project implementation was often dominated by the problems of technology installation and operation, diverting attention from the softer (yet critical) issues of technology assimilation by the community. When working with rural communities in developing countries, it is not possible to make assumptions about pre-existing levels of knowledge or awareness, either about technology or about development. Practitioners need to work beyond the technology of telecentres, difficult as that often is, and to engage closely with the dynamics of the community if they are to ensure success in terms of development.

In many ways, telecentres focus more attention on the processes of development, especially those concerned with local communities, than they do on the technology that they promote. The lesson seems to be if you want to know how a community can use technology, study the community rather than the technology. This approach implies the use of interpretative methods, which prefer meaning over measurement and where data collection and representation have been accomplished with informant interviewing ethnography, or the thick description of cultures based on intimate knowledge and participation, and even ethnographically linked textual analyses (Spradley, 1979; Van Maanen, 1988; Gephart, 1993).

It is also worth noting that communities in the developing world are not adopting ICTs out of their own volition. Where they are in use, this has usually been as a result of outside action, either by development agencies, researchers or governments. Accordingly, the telecentre success model emphasizes the social dynamics of communities, which determine the extent to which they will embrace and share in the development goals that might be externally imposed and which the telecentre makes possible.

Telecentres offer a solution to the digital divide and they present significant development opportunities for the vast majority of people living in developing countries. Understanding what makes telecentres successful is crucial if they are to fulfil this promise. The case studies provide an insight into the factors that lead to telecentre success. A general theory of telecentre success is urgently required. The time is now right to stake stock of the experience available from pilot projects and to begin distinguishing between what is working and what is not. The findings suggest some practical lessons for development.

Telecentre and information characteristics are relatively mechanical in that a telecentre has to be well managed and it has to provide information to its community that will be put to good use. Mechanisms for accomplishing these outcomes are reasonably prescriptive. Suitable methods for telecentre management are relatively simple and universally applicable. Techniques for eliciting information needs may be a little more demanding, as this requires intimate knowledge of the user community. The structural conditions surrounding telecentre innovations are likely to be less tractable, being subject to external influences and agencies. However, the characteristics of the community that uses the telecentre are much less likely to be subject to influence by the activities of telecentre staff, management or promoters.

Community characteristics could be the most potent factor in influencing telecentre success but they are also the least manageable. Moreover, due to their structural conditions, all the projects described here required a variety of innovative technical solutions involving wireless Internet access and electricity provision requiring diesel generators, solar panels and battery power. The challenges of putting this together in less than perfect surroundings, whilst wholly necessary, tends to divert project officers from other critical path factors such as community characteristics.

As stories begin to emerge of the development successes that some pilot telecentres have achieved, there is a dangerous impression that marginalized communities need only be given access to technology in order to share in the benefits of the information age. The findings from the cases here suggest that there is as much or more about the community than there is about the technology that determines telecentre success. The community characteristics variables are arguably the most potent in predicting telecentre success. Experience indicates that whilst technology is necessary for the transformational opportunities that are available, of itself, technology is insufficient. A model of telecentre success must reveal what else is required for ICTs to achieve their fullest potential in development situations.

Acknowledgements

This study was supported by research grants from the International Development Research Centre (IDRC) of the Canadian Government. The author acknowledges the assistance of Professor M. Lynne Markus for guidance and comments on an earlier draft and acknowledges the comments by other reviewers.

References

Benjamin P. (2000) Telecentre 2000, Link Centre, Witwatersrand University, Pretoria, Republic of South Africa, www.communitysa.org.za/tele2000.htm.

Colle, R. (2000) Communication Shops and Telecentres in Developing Countries, in: Gurstein, M. (ed.) *Community Informatics: Enabling Communities with Information and Communications Technologies*. Hershey: Idea Group Publishing.

FOOD, Foundation of Occupational Development, Chennai, Tamil Nadu, India: www.xlweb.com/food/wireless.

Gardner K. and Lewis, D. (1996) *Anthropology, Development and the Post-Modern Challenge*. London: Pluto Press.

Gephart, R. P. (1993) The Textual Approach: Risk and Blame in Disaster Sense Making, *Academy of Management Journal*, 36, 6, 1465–1514.

Harris, R. W., Bala, P., Songan, P. and Khoo E. (2001) Challenges and Opportunities in Introducing Information and Communication Technologies to the Kelabit Community of North Central Borneo, *New Media and Society*, 3, 3, 270–295

Hudson, H. E. (2001) Telecentre Evaluation: Issues and Strategies, in: Latchem C. and Walker, D. (eds.) *Telecentres: Case Studies and Key Issues*. www.col.org/telecentres

Kanfi, S. and Tulus F. (1998) Acacia: What Is A Telecentre? British Council, Johannesburg, Republic of South Africa. web.idrc.ca/en/ev-6646-201-1-DO_TOPIC.html

Markus, M. L. and Soh, C. (2002) Structural Influences on Global E-Commerce Activity, *Journal of Global Information Management*, 10, 1, 5–12.

MSSRF, MS Swaminathan Research Foundation (2000) Assessment of the Impact of Information Technology on Rural Areas of India. Terminal Report for the IDRC.

PANTLEG (1999a) Success Stories of Rural ICTs in a Developing Country. International Development Research Center, Canada, www.idrc.ca/pan/telsuccsttories_e.htm.

PANTLEG (1999b) PANAsia Telecentre Learning and Evaluation Group (PANTLEG), Research Support Project #100579, IDRC, www.idrc.ca/pan/pr100579_e.htm.

Roman R. and Colle R. D. (2002) Themes and Issues in Telecentre Sustainability, Development Informatics Working Paper Series: 10, Institute for Development Policy and Management, University of Manchester, www.man.ac.uk/IDPM

Spradley, J. P. (1979) *The Ethnographic Interview*. New York: Holt, Rinehart and Winston.

Songan, P., Harris, R. W., Bala, P. and Khoo E. (2000) Awareness and Usage of Information Technology in a Rural Community of Bario, Sarawak, *Proceedings of the Sixth Biennial Borneo Research Conference*, 10–14 July, Kuching, Sarawak, Malaysia: 626–648.

Van Maanen, J. (1981) *Fieldwork on the Beat: An Informal Introduction to Organizational Ethnography, Conference on Innovations in Methodology for Organization Research*, American Psychological Association.

Whyte, A. (1999) Understanding the Role of Community Telecentres in Development—A Proposed Approach to Evaluation, in: Gomez, R., and Hunt, P. (eds.) *Telecentre Evaluation: A Global Perspective*. www.idrc.ca/telecentre/evaluation/nn/00_Cov.html .

World Bank (1999) *World Bank Development Report, Knowledge for Development*. The World Bank, Washington DC.

About the Author

Dr. Harris has been advancing the use of ICTs for development since 1997. He is a specialist in rural access to ICTs, having conceived, instigated and lead for its first three years the pioneering e-Bario Telecentre Project in Malaysia. This project received multiple international awards, as well as an award from the Prime Minister.

Dr. Harris has provided consulting services internationally to Asian governments and aid agencies, including the UN and World Bank. In 2002–03, he spent several months in Nepal working with UN agencies developing projects to provide rural access to ICTs for poverty alleviation. This consisted of establishing a pilot project with 15 rural telecentres in remote locations, capable of delivering useful development information to the local communities. In another case, an existing programme was working with rural-urban linkages involving 12 locations and was seeking to empower peri-urban communities with ICTs for purposes of e-government and e-commerce, which was being piloted for wider implementation nationwide. Additionally, an existing project that had been conducting a programme for social mobilisation in 23 rural villages was planning to install telecentres in order to energize the programme towards higher levels of localized development achievements. Dr. Harris conducted assignments in Vietnam with UN agencies in support of policy regarding ICTs for poverty reduction as well as pilot rural telecentres. One project facilitates VietNam's accession to World Trade Organisation through rural ICT projects that are trade-oriented, in order to familiarize Vietnamese policy-makers with the potential for ICTs to invigorate the rural economy by providing access to international markets and improved sources of production related information.

In other instances, Dr. Harris worked with the World Bank to assist in the specification of a rural telecentre programme as part of the e-Sri Lanka programme of the Government of Sri Lanka, and with UNDP and the Indonesian Government's national planning agency to develop a national programme for poverty alleviation with ICTs in Indonesia. More recently, Dr. Harris has been involved with discussions and proposals concerning forthcoming pilot projects to test the efficacy of rural ICTs in the promotion of pro-poor, community-based tourism. These have involved the Asian Development Bank in the Greater Mekong Sub-region of Laos, Cambodia and Vietnam, the Dutch national aid organization (SNV), also in Vietnam, and with the Tourism and Trade Department of the East Malaysian state of Sabah.

Between these consulting assignments, Dr. Harris has provided knowledge resources to a variety of international capacity-building exercises, including workshops for UNESCAP, UNDP-APDIP, UNDP China, UNESCO, APEC, ITU and FAO. Other assignments have been concerned with the evaluation of poverty-reducing ICT initiatives; in India for UNDP-APDIP and in Mongolia, Philippines, India and Malaysia for IDRC (Canada).

6. Sustainability Issues in Rural Telecentres—
An Overview with Two Case Studies

V. Balaji

International Crops Research Institute for the Semi-Arid Tropics (ICRISAT)
Patancheru 502324, AP
India
V.Balaji@cgiar.org

1. Introduction

The word "telecentre" is not found in a standard dictionary of the English language, but it brings up a number of images, the dominant one being that of a "cyber café" found in the urban areas of Asia or Latin America. Although the concept of rural telecentres is frequently discussed, a generic definition remains elusive. For example, is a "public call office" in a rural setting, containing only a telephone, a telecentre? Or, does the well-known case of the use of mobile phones in semi-rural settings in Bangladesh represent rural telecentre operations? Going by the definition proposed by Hudson (1999) that there must be something "tele" in a telecentre, these should be viewed as sound examples of rural telecentres. The inclusion of an information function in a telecentre, however, introduces an important new dimension among the prevalent notions of a telecentre, which relate more closely to its communication aspects. It is assumed that the information function is taken care of once the communication function is fulfilled.

In this chapter, we will point out that a rural telecentre must place simultaneous emphasis on both the communication and information functions and that the core issue of sustainability of a rural telecentre is dependent upon adopting this perspective. To clarify some of these issues, we shall present two case studies from India where, according to Keniston (2004), a large number of pilot rural telecentre projects have been launched (or shut down) in the last five years.

2. Working Definitions

The definition of the word *"rural"* in the development research literature varies widely. The majority of statements use production systems as their basis, while a number of poverty-related definitions of rural are also current. For our purposes, we shall adopt a working definition— a locality will be considered "rural" if the majority

of the livelihoods of the humans there are dependent upon harnessing natural resources (soil, water, biological diversity, for example) in a primary way. A further
consideration will be based on how far removed a locality is in terms of access to information "markets" or information sources and the exchange processes associated
with them. The former would be considered a necessary condition whereas the latter
will be a supplementary criterion.

Given such a perspective, a *rural telecentre* will necessarily provide information
services as much as providing at least one of the components of contemporary communication facilities. The information services that are the integral component of a
rural telecentre will depend upon the type of communication facility offered as well as
on the local demand patterns.

The core issue of *sustainability of operations* arises from the ability of a rural
telecentre to meet and fulfil local demands for information services and communication facilities. This is treated as a given. Nevertheless, there are different dimensions
of sustainability that need to be stated before a substantial understanding of the issue
can be developed. Three such dimensions are the *technical*, the *financial* and the *programmatic*. The technical relates to both the communication and the information service aspects; connectivity, for example; or, databases/interfaces design using local language typefaces; or even licensing issues. The financial relates to investments, revenue
and expenditure. The programmatic aspect relates more directly to the information
services; examples include value addition to generic information to generate locally
relevant knowledge; training of the telecentre operators and at least a section of the
users. It is sufficient to note that unsustainable practices in any one dimension can
result in the operation as a whole becoming unsustainable.

3. Information Services in Rural Telecentres

The heart of a rural telecentre's operations is in the information services offered to its
clientele. Traditionally, there have been a number of information channels, some
formal, most non-formal or informal, that have served rural families, even if only in-
adequately. These include agricultural extension services, community health projects,
village reading rooms, community meetings, religious places, etc. Broadcast media
such as the radio and the television are important information channels, while newspapers and magazines, with their limited reach because of prevalent non-literacy, are
also known to be significant channels of information in rural areas. The advent of the
Internet and digital information and communication technologies (ICTs) has introduced unprecedented opportunities to enhance the information channels available in
a typical rural area in the developing world, with the possibility of large-scale improvements in intra-rural and rural-central information flows.

The rural telecentre is thus a facility that enhances intra-rural and rural-urban/metropolitan information flows using some of the contemporary ICTs. From such a point of view, a rural telecentre is not just a facility with a network terminal device, but provides a variety of information services that are locally in demand. It should fulfil at least some of the information services available from the more traditional channels even as it offers altogether new ones.

The rural telecentre thus connects to contemporary information networks on one side, so to speak, while connecting to a rural user on the other. Although it appears at a first glance to function like a relay or a switch, it has important additional functions. The most important of them is one of value addition. A rural user, with relatively limited exposure to contemporary computing and communication devices, will not find the rural telecentre useful unless the generic information (found in the networks) is delivered to her as local information that she can act on. Even a simple step like translation from English into the local language is a value addition, and many more examples can be given. In agriculture, network information is always found to be too generic to be acted on locally in a rural setting and requires much transformation before it is ready for use. Similar is the case with information on basic hygiene and public health.

Value addition is therefore a significant function in a rural telecentre and this recognition changes the techno-centric character of many a telecentre. There have been statements that instead of bringing frontier technology, adoption of such a perspective would bring only information with relatively reduced emphasis on connectivity or communication technologies. However, as we shall see later, putting the rural user in a priority position makes the rural telecentre considerably more sustainable programmatically, even if it reduces the charm of introducing frontline technologies among the "unreached" population.

4. Value Addition, Intermediation and Capacity Building

Individuals carry out the process of value addition and this necessarily raises the issues of intermediation. The advent of the Internet and the rise of E-Commerce in some parts of the world were hailed as pioneering "disintermediation" generating greater value to the actual user or consumer. Here is a case involving the Internet where we are claiming that intermediation is associated with value addition. We believe there is no contradiction here because there is no perceived value chain in rural information flows where the intermediaries are appropriating part of the value. The rural information chain, if we can use the name for a collection of channels, is not a value chain largely because few information products in the organized production and marketing sector are meant for the rural users primarily (movies in South Asia used to be an exception). The presence of intermediaries in a rural telecentre will eventually lead to creation of a value chain of a novel kind because of the way it would deliver new services, both conventional and novel, to rural users. A few examples will be described in the case studies below.

The presence of intermediaries in the rural telecentre and the functions of intermediation bring up other related issues that pertain to power arrangements in rural areas. The variety of iniquitous relationships conditioned by the local economy, culture, social arrangements (caste or untouchability in rural India, for example) and gender have a significant influence on the way in which information intermediation is carried out. In designing and developing intermediate functions, therefore, more factors than just technology inevitably come in, and cannot be avoided. The process of info-mediation in a rural telecentre thus should not only be relevant and demand-driven, it should also be socially and culturally credible to attract and retain a large user base. The adoption of such a perspective takes the rural telecentre even further away from a technology-centric perspective.

5. Technological Issues in a Rural Telecentre

The 1990s saw the rapid development and equally rapid spread of a number of technologies in the spheres of computing and communication that simplified access and made network access affordable in most Organization for Economic Co-operation and Development (OECD) countries. The issues and the costs of connectivity, however, have received a disproportionate share of attention in the debates on technology issues that concern the establishment and operations of a rural telecentre. The debate acquired a strongly technology flavour primarily because the issues of connectivity were given prime importance. The technology for content creation and access to the content has by and large remained in the shadows. The connectivity-first approach in many initiatives has led to a situation where the financial and management burdens overwhelm the proponents of rural telecentres resulting in poorer value addition to the essential information products and services that rural people require. With a number of connectivity technologies competing for the limited resources of rural telecentre proponents, it can only be said that the perspective should firmly be anchored in a rural-user-first approach.

The core of the local content-creation process is in the standardization of typefaces, which is not yet a reality in most languages in the developing world. On occasions, the power of corporate monopoly has led to standardization of typefaces in a small number of languages as an unintended consequence, but most languages in Asia and Sub Saharan Africa do not have this advantage. This has led to situations where the proponents as well as the local operators in rural areas need to be conversant in English or French in order to create and store content in a retrievable format. This is one area that still needs investment from actors in the development research arena, and needs even more policy support and guidance from governments in many linguistic regions. The significance of multi-lateral, non-governmental organizations such as the Unicode is evident here.

The issue of connectivity is very largely related to cost, licensing procedures and local capacity for maintenance and support. Although the costs have been declining, affordability is still not possible as far as rural areas of the developing countries are concerned. The licensing procedures for the use of terrestrial wireless connectivity technologies remain cumbersome and difficult in many countries while direct access to satellites is allowed only in a small number of countries. Most of the wireless data networking technologies are known to impose huge hidden costs in maintenance. The result is that proponents of rural telecentres are still unable to identify a generic solution to the connectivity issue, and the most practical option appears to be to work on limited capability, low-cost solutions in the medium term. There is a need to strike a balance between the charm of introducing the latest in network technology in a rural setting and the affordability and sustenance of a rural information enterprise.

6. Case Studies in Operational Rural Telecentres

We shall now look at two cases where rural telecentres have been launched and sustained for a period of time: a series of micro-level studies in the process of consolidation and publication in 2003 in relation to India; and another set of studies covering different parts of Asia that have been published on the web (www.rogharris.org). According to Keniston (2004), there is still a need to undertake a rigorous and systematic study of the rural telecentres; their operations, costs, and impact, across a wide spectrum of locations and actors. We offer these two case studies out of both published information as well as fresh data gathered from a recently launched rural telecentre project. Both are located in India.

6.1 Village Knowledge Centres Project in Pondicherry, India

This is a project initiated by a research-oriented non-profit organization in India and details may be found at www.mssrf.org. The project was established in 1998 to develop a method to assess the impact of ICT on rural development. The region was chosen because the level of human development was better than the average for India. The region is predominantly agriculture-dependent with an urban centre and road network with about one-fourth of the population living at a daily income level less than one US Dollar a day. Tamil is the principal language spoken in this region.

The project started with setting up an access centre in one of the larger villages that was connected to long-distance calling facilities and had access to a 2.4 KBPS dial up network to the Internet. This office was connected to village access points using a full-duplex VHF radio network (open to the Indian public for use under low transmission power) that was capable of voice and data transmission. The village access points were set up as community centres in places chosen by the village community and operated by volunteers elected by the village. They were given basic training in PC operations and in the use of the newly developed Unicode-compatible Tamil typefaces and fonts. With the help of this group of volunteers, an appraisal of the prevalent information flows and channels in the locality, as perceived by the residents, was carried out. This appraisal revealed that most rural residents, who are information-poor in our sense, were linked primarily to others like themselves; the major external sources of information were the radio and the TV channels. Interestingly, the local merchants were also important sources of useful information (see Figure 6.1).

Figure 6.1
Information Flows in the Rural Areas of Pondicherry, India, as Perceived by Rural Residents (from Balaji *et al.*, 2000)

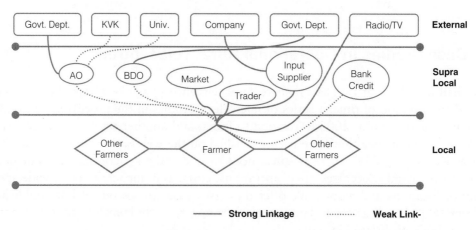

AO; Agricultural Officer; BDO: Block Development Officer; KVK: Local Farm Science Centre

During the initial phase, the project proponents attempted to introduce substantial content relating to agronomic practices. The utilization of such information was not high. However, a series of consultations with various segments of the rural residents there revealed that such information would be of interest only to a relatively small and affluent group who did not need an additional channel. This group was well connected in the information flows starting at the local government level, which served them well. Agriculture was the predominant mode of generating local livelihoods here, and the primary demand for information was for market prices in the supra-locality (approximately a distance of 50 km). The other demands related to obtaining unambiguous and dependable information on local government projects to benefit the rural poor. Quite significantly, education-related information (about high school or college courses in the larger locality) emerged as an important need (Table 6.1).

Table 6.1

Pattern of Usage (in %) in the Pondicherry Village Centres (up to June 2000)

Voice	(%)
Personal	16.33
Programme related	1.18
Data	
Agriculture & Fisheries	6.69
Education & Training	32.01
Employment	2.05
Health	1.51
Govt. Sector / Entitlements	40.43

Source: Balaji et al. 2000: a total of 15,000 "hits" from the registers in three village telecentres were analyzed for this purpose. The period covered is 18 months to June 2000.

These results changed the way the access point at the bigger village was functioning. Instead of acting only as a gateway to the Internet, this centre became a "switch" that blended information from a variety of local and network sources and delivered it via the local VHF network. This is an example of value addition to generic and dispersed information. The fact that the village centres have endured for over three years with no cash incentive from the project shows that an important demand in rural development was being fulfilled using value-added information. The access to the Internet, even if limited to just one centre proved to be extremely useful to another village on the VHF network that is totally dependent on artisan fishing in the Bay of Bengal. This village centre received, via the value-adding centre, regular predictions on wave height changes in the area downloaded from the web site of the US Navy at Rota, Spain. These predictions were considered accurate and life saving by the fishers. The predictions were rendered in Tamil, the local language, and announced using loud speakers kept in the village centre.

Thus, what was important was to set up a local network that is useful in delivering value-added, locally relevant and "actionable" information in the right time. Connectivity is important but not in a "traditional" sense. Connectivity is enhanced in its value if locally specific information can be obtained faster. This is also exemplified in the case of the fishing village, which could not have obtained the wave-height prediction without access to contemporary ICT. The project did not invest heavily in a connectivity infrastructure but instead focused more on assessing the demand for information and delivering it with quality and in a timely fashion. Considerable effort was made to build capacity among the volunteers in not just PC operations, but in assessing local demands for information and in timely delivery.

The role of credible intermediaries in unfolding and sustaining new information flows emerges clearly in this case. The project did not develop a fee-payment arrangement to support the local telecentre services, but fostered a method by which the village assembly voted an in-kind contribution (rent-free space of good quality, free electricity, free time of village youth as volunteers) using a legally viable memorandum that was renewed ever year. The project was sustainable from the programmatic and technological points of view while in a strict sense of the term financial, it has not developed a revenue-based approach but has developed a unique cost-sharing arrangement.

6.2 A Rural Information Hub for Watershed Management and Drought Mitigation in Andhra Pradesh, India

The Adarsha Welfare Society (AWS) is a federation of rural, all-women micro-credit groups in the Addakal Mandal of the State of Andhra Pradesh in India. The total membership of the Society is about 4,500, covering 37 villages in the vicinity (total population about 45,200). The Society established its own training centre out of its savings and with help from the local development funds with a market outlet plus a highway restaurant.

This area or rural cluster is one of the poorest regions in India. There are pockets here that experience a consistent and severe scarcity of water. The year 2002 was a particularly bad year because the entire State was facing a serious drought; there was an out migration of population, mostly male workers, which placed enormous burdens on the womenfolk to sustain the family. For a long period of time, animal husbandry has been an important source of livelihood but with continuing drought conditions semi-nomadic practices involving small animals (goats, for example) are becoming important. The AWS believed that the best way of coping with drought was to be able to anticipate it and prepare for it. A number of agricultural research institutes, including the ICRISAT, which is an international agricultural research centre, joined the AWS and a State-supported project to establish an information system on drought. The AWS offered to host it in its marketing-cum-training centre premises.

Six women and men identified by the AWS were trained in all the basic PC operations and in the extensive use of the Telugu language typefaces that are Unicode-compatible. Although a telephone was available in the premises, it was not found to be "stable" and could not hold an Internet connection for longer than ten minutes. With the support of the State agency, low cost direct satellite connectivity to the Web was set up (legal in India since late 2001). The trainees acquired sufficient skills in browsing the web for information and helped organize a participatory appraisal of the information flows in the locality (covering all the 37 villages). The outcome of the appraisal is presented in Figure 6.2.

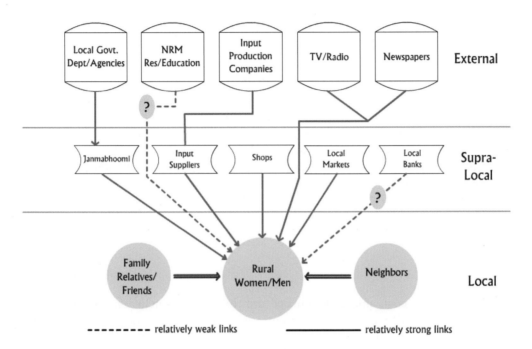

Figure 6.2
Information Flow and Linkages in Addakal Mandal (AP, India)
as perceived by Rural Women and Men (May, 2003)

What emerges is the very important role that the local retailing arrangements play in information flows. The agricultural extension system is not rated highly as an information source while the local government receives a high rating. The media combination is also rated well. An assessment of information demand revealed that the primary need is for market information and on wages for labour. There is also considerable demand for information on new opportunities for education. Weather information was not in particularly high demand. Although the AWS is a federation of micro-credit societies, there was no discernible interest in using a computing platform to conduct their business except to store some records.

Based on this analysis, the AWS-ICRISAT approach is to build an information retailing system rather than a warehouse so that the cost of information inventory is reduced. The emerging telecentre focuses on providing information in relation to the local commodity and wage labour markets, and on resource education under the umbrella of a Virtual University, a coalition of national and international agricultural research centres and Open Universities (Navarro and Balaji, 2003), with water management and drought mitigation as key interests. The first collection of modules were developed and tested with the membership of the AWS over a period of six months. The responses revealed that there is very little knowledge about the causes, extent and impact of droughts, and there is a strongly expressed need for learning more about them. Market price information is now made available on a daily basis at the hub, and it is available on local notice boards in the Telugu language. Important news, downloaded from newspaper web sites, is summarized and made available in Telugu on notice boards.

This case study again shows the importance of building up a comprehensive knowledge base on local information channels and popular perceptions about which sources are useful. The operations are considerably dependent upon such knowledge. The programmatic sustainability is thus built in. In this case, an easily licensed and serviceable technology for connectivity is used and the typeface problem has been addressed ab initio. Financial sustainability is still an open issue; the cost of content creation (learning/educational materials) is high and this is not currently borne by the local organization although they are bearing the operational costs. This is partly addressed in the way a virtual group of experts, spanning a number of institutions, generate meta-content which can be adopted by a local organization at a relatively low cost.

7. Conclusion

The rural telecentre is an emerging concept and the word "conclusion" is not wholly appropriate here. Until recently, the telecentre movement was preoccupied with technology (connectivity, that is) issues, and only now are issues of equity and content creation being addressed. The sustainability of a rural telecentre is dependent upon how it fulfils local demand for information, even as it adds novel and useful information channels using contemporary ICTs. The technology choice should be based primarily on practical considerations, rather than on its frontline nature. Financial sustainability still remains a key concern. Although many workers have advocated a revenue model for financial sustainability, it is clear that the costs of connectivity (capital plus recurring costs) and of content generation can quickly overwhelm a rural telecentre where local disposable income is meagre or non-existent. One strategy is to foster cost-sharing arrangements with the rural community contributing in kind, or to make use of rural infrastructure developed by local organizations. An interesting development is the possibility of forming virtual groups of experts who can interact with telecentre proponents for special purpose information. Thus, a rural telecentre will eventually be one that would mediate between virtual groups of experts and network information sources and rural users who require specific and locally actionable information. It would be a key component in an emerging information market that has products for rural clients.

References

Balaji, V., Rajamohan, K. G., Rajasekarapandy, R. and Senthilkumaran, S. (2000) Towards a Knowledge System for Food Security: The Information Village Experiment in Pondicherry, *OnTheInternet*, 6, 2, 32–37

Hudson, H. (1999) Designing Research for Telecenter Evaluation, in: Hunt, P. and Gomez, R. (eds.) *Telecenter Evaluation: A Global Perspective*, web.idrc.ca/uploads/user-S/10244248430Farhills.pdf.

Keniston, K. (2004) Introduction, in: Keniston, K. and Kumar, D. (eds.) *Equity, Diversity and IT*. New Delhi: Sage Publications.

Navarro, R. and Balaji, V. (2003) A Virtual University for the Semi-Arid Tropics: A Concept Note, *Electronic Journal of Information Systems in Developing Countries*, 14, 5, 1–10.

7. Towards Demand-Driven Community Telecentres

Royal D. Colle

Professor Emeritus
Communication Department
Cornell University
rdc4@cornell.edu

Raul Roman

Research Associate
Center for Internet Studies
University of Washington
rroman@u.washington.edu

Abstract

Telecentres have become a major part of the global movement to use information and communication technologies for development, the authors concentrate on major factors that influence the demand for telecentre services. Most of these apply in some significant way to the quality of information and communication services that telecentres provide for their host communities, or to the motivation of communities to patronize telecentres. This chapter explores such issues as university involvement in telecentre initiatives, gender issues, the role of research, and the collaborative environment of telecentres.

1. Introduction

Here are two pictures related to the spread of Information and Communication Technologies (ICTs) in Asia. First, to South Korea:

> Burrow a bit down the alleys, up flights of stairs, or into the corners of malls, and you find something that sets Seoul apart and fosters its passion for broadband—online game rooms, or PC *baangs*, as they are called there. There are 26,000 of them, tucked into every spare sliver of real estate. Filled with late-model Personal Computers packed tightly in rows, these rabbit warrens of high-bandwidth connectivity are where young adults gather to play games, video-chat, hang out and hook-up. They are known as "third places"—not home, not work—where teens and twenty-

somethings go to socialize, to be part of a group in a culture where group interaction is overwhelmingly important (Herz, 2002).

Now to South India, and to a small kiosk where we were able to send an email to North America using a PC and the latest version of wireless networking technology. There were no other users in sight and the owner guessed that he might have five persons step in during a typical day. When we asked why so few, he suggested that people just did not know about the place.

While the Korean *baangs* are supported by consumer fees and sales of refreshments adding up to a US$6 billion business in 2001, the Indian enterprise, substantially subsidized by a private sector agricultural conglomerate, seemed hardly able to survive. While there are vast differences in the two situations that make significant comparisons inappropriate, we do see one case where demand is clearly evident and one where demand seemingly does not exist. Demand is a complex issue especially in situations where consumer and community participation are important in measuring success and determining survival of enterprises; and demand is a vital part of the emergence and sustainability of the kinds of ICT community initiatives that are keys to developing countries becoming active players in the Information Society.

In this chapter, we focus on issues related to "demand-driven telecentres", defining these simply as telecentres that offer services and resources that are wanted and used by many in the host community. This contrasts with initiatives that are introduced without careful assessment of the community's desires and without appropriate consideration for exciting the community about an innovation. These imposed or uninvited innovations litter the field of development. We will explore some of the factors that influence demand for telecentres in developing countries particularly outside their urban cities. Our exploration will have several twists and turns. We start with a brief look at communication and community development.

2. Information and Communication Technologies in Development

2.1 Major Thrusts in Development and Communication

In a recent United Nations Educational, Scientific and Cultural Organization (UNESCO) publication, we identified some of the major threads in the evolution of development communication (Colle, 2003). These were not theoretical or philosophical threads; they were major action-oriented approaches in which communication played significant roles in development programs. We describe each one briefly.

2.1.1 Population and Communication

The acronym IEC—Information, Education and Communication—has achieved greatest prominence in programs designed to influence knowledge, motivation and behaviour related to contraception and family planning. National governments, NGOs, multi-national agencies, and the private sector have conducted many studies and interventions in which population issues and communication and have been central components. These programs, through their successes and failures, have enriched development communication through their practically oriented explorations in media use and message innovations. The UN Population Fund was among the first to use the term IEC in 1969, when it used the label for its population communication activities that were largely concerned with how it could most effectively persuade people (particularly women) to adopt new birth control methods. Since the 1994 International Conference on Population and Development in Cairo that helped turn family planning into "reproductive health", IEC has acquired a broader mandate, including an emphasis on advocacy and gender issues. These reinforced the placement of information and communication technologies and the newer digital communication tools on the intervention agenda.

2.1.2 Social Marketing and Behaviour Change

Social marketing has become a dominant approach in promoting behavioural change in development initiatives. Although ICTs are not inherent in the social marketing approach, the strong emphasis in social marketing for reaching out to a large number of people, and on reaching them with frequent repetition of messages, inevitably links its strategies to the mass media. Social marketing imitates many of the features of commercial marketing where mass media advertising and an emphasis on "reach and frequency" play a major role (Andresen, 1995)

2.1.3 The Entertainment-Educate Approach

The discovery that entertainment can be harnessed for increasing the effectiveness of social change interventions has led to an emphasis on using music, dramas, and a variety of other formats to promote causes ranging from women's rights to automobile safety. Although instruments such as street theatre illustrate this approach, ICTs, including radio, television, and CDs, have become prominent carriers of entertainment for education. In India, *Tinka, Tinka, Sukh* (Happiness Lies in Small Things) on radio, *Hum Log* (We People) on television and the multi-media *Soul City* project in South Africa represent how "enter-educate" has been used for social and behavioural change across the world (Singhal and Rogers, 1999).

These parallel thrusts that continued into this century remind us that conventional media such as radio, television and recordings are still important members of the ICT family. The enormous advances in audio and video technology that have lowered the costs, simplified operations, and miniaturized the communication tools reinforce this point. In fact, some argue "the telephone and radio might have a higher benefit-cost ratio and lower overall cost as alternatives to and intermediaries for the Internet in poverty alleviation programs" (Kenny, 2002).

The various initiatives that have taken place using these approaches point out the significant role that ICTs, broadly defined, are playing in development programs. As contrasted with approaches that emphasize face-to-face and small group methods of contact, ICTs offer some significant benefits where information and communication are important factors. These include the following:

- Reaching many people simultaneously,
- Overcoming geographical boundaries,
- Overcoming social and literacy barriers,
- Providing frequency and repetition of contact,
- Storage of information for on-demand access,
- Capturing the reality of events,
- Greater efficiency (lower costs) in sending and receiving information.

In the 1990s, computers and digital networks exploded into the communication environment and provided additional dimensions to the ICT benefits inventory. These newer technologies provided at least four additional features. These included opportunities for:

- Relatively convenient individual information searching through a vast array of information sources, often 24 hours a day;
- Timely interaction between and among computer users that allows convenient and "contemplated" exchanges—exchanges that are quick but not necessarily instantaneous;
- "Broadcasting" of information to many by ordinary individuals, including easier "bottom-up" message initiation;
- Global reach almost constantly.

These features of ICTs can translate into benefits in education and health, reducing social distance, better connections between governments and individuals, marketing advantages, and, overall, improved opportunities for information sharing. However, there is still a more profound implication regarding ICTs. According to some analysts, writing and reading have long been symbols of, and contributors to, social inequality because these skills provide their holders with information and knowledge that lead to power and privilege. Now, say some, "the rise of multimedia should provide an important opportunity to level the playing field of literacy by restoring the status of more natural forms of audiovisual communication that are in some ways more broadly accessible" (Warschauer, 2003). However, widespread discussion in this decade about the "digital divide" points to a concern about the possible negative consequences of the uneven access to computers and telecommunications networks among people within nations and, more broadly, the gap among nations themselves.

2.2 The International Advocates

At a 2002 Special Session of the UN General Assembly on information technology and development (17–18 June), Secretary General Kofi Annan succinctly placed ICTs among the important tools of development. Opening the session, he declared:

"A wide consensus has emerged on the potential of information and communications technologies to promote economic growth, combat poverty, and facilitate the integration of developing countries into the global economy. Seizing the opportunities of the digital revolution is one of the most pressing challenges we face".

The potential of ICTs for development, highlighted by computers and networks, has prompted countless international conferences and workshops in an attempt to harness this potential. Enormous hours of work and untold sums of money were poured into the preparations and arrangements for the World Summit on the Information Society (Geneva 2003 and Tunisia 2005) in part to understand the implications of what has become known as "The Information Society". The major actors include both multi-national and national bodies such as the World Bank Group, UNDP, ITU, UNESCO, FAO, USAID, IDRC and others.

The widespread belief that computers, networks and other kinds of information and communication technologies can positively affect development has led to various approaches to providing people with the use of these resources. In some countries, the emphasis has been on developing low cost computers such as India's *Simputer* and Brazil's *computador popular*, and promoting lower cost connectivity such as wireless local telecommunication loops (Warschauer, 2003). However, the cost of individual ownership and connectivity remains well beyond a majority of the population in developing nations.

3. Telecentres

3.1 The Emergence of Telecentres

To address the challenge in nations from Mexico and Brazil in Latin America to the governorates in Egypt and the villages of India and Nepal, a wide range of organizations have promoted and supported the creation of local entities that make ICTs available on an affordable basis to many who could not afford them individually. Much of the attention is on computers and telecommunications networks, and follows an approach in countries such as India where community viewing centres were for many years the way in which the population watched television. The idea of a community sharing computer technology emerged prominently in the 1980s (before the Internet) especially with the introduction of the telecottage in Scandinavia. The initial purpose of those telecottages was to fight against the marginalization of remote places by providing rural people with the new tools of the emerging Information Society.

With the sprouting of the Internet in the 1990s, a new breed of public access appeared. The telehouses in Hungary are an example. Supported initially by the US Agency for International Development (USAID), these telehouses offered computer and Internet access, but especially concentrated on the social and economic development of rural communities. Oddly, the Hungarian program was less an ICT initiative and more an effort to help Hungary revitalize local government after the collapse of the nation's centralized political and economic structure (Gáspár, 2001; Bihari and Jókay, 1999).

Hungarian telehouses were part of the robust movement that marked the close of the 20th century, with a momentum around the world that has continued into this century. The movement includes the emergence of three major types of community ICT-related shared facilities—telecentres, cybercafés and information access points (IAPs).

3.1.1 Telecentres

Telecentres tend to be in the public sector, operated by governmental bodies or non-governmental organizations (NGOs). Generally they serve a low-income clientele, and have a community development mission. Typically, telecentres offer a broad range of communication services related to the needs of the community, some of which are free or subsidized by external bodies such as governments or NGOs. Along with computer and Internet access, these services might include—desktop publishing, community newspapers, sales or rental of audio and video recordings, book lending, training, photocopying, faxing, and telephone services. Some, like the Hungarian telehouses and Western Australia's rural telecentres, provide postal, banking and employment services.

3.1.2 Cybercafés

The commercially-oriented cybercafés that are found from the streets adjoining China's Tiananmen Square to the neighbourhoods of Buenos Aires have been an equally energetic movement. They are usually in the private sector and focus primarily on providing customers with the use of computers and connections to the Internet and the Worldwide Web. Their clients tend to be more urban, more educated, and more economically well off than the clients of telecentres. The principal attractions at the cybercafés are computer games and e-mail. For example, the *baangs* in Korea illustrate this widespread application of the "shared facilities" approach in the private sector. Similarly, in Peru, almost 3,000 private sector Internet *cabinas* are sustained by charging fees to users (about US$0.40 per hour) for PC, e-mail and Internet services. In Latin America, the private sector and NGOs are particularly active in supporting public access initiatives.

3.1.3 Information Access Points (IAPs)

Information access points fall between the cybercafé and telecentre approaches. As they focus on Internet and network services, they emphasize the opportunity for the community to seek information. The most dramatic example is Canada's Community Access Program that established 10,000 access points in rural and urban areas across the country between 1994 and 2001. Computers and network connections were placed in community centres, libraries, schools, and other public places in order to make Canada "the most interconnected country in the world" (Pfiester *et al.*, 2000). Canada's success energized other national IAP initiatives: in 2002, the Government of México designed a network of *Centros Comunitarios Digitales* (DCCs) as part of its *Sistema Nacional e-México*. The Government's plan is to have more than 12,000 DCCs by 2006, covering 75% of the nation's population (Rodriguez Oliveros, 2002). In the Indian State of Tamil Nadu, the project called Sustainable Access in Rural India (SARI) initiated a plan to establish "tele-kiosks" in up to 100 villages in Madurai District as the first phase of an initiative that will see thousands of IAP kiosks flooding villages all over the state.

IAPs sometimes are reinforced by sectoral organizations such as those in health and agriculture that build special information systems. An example is the World Health Organization, which has set out a seven-year plan to establish the Health InterNetwork Project. It is an initiative to facilitate the flow of health information worldwide using Internet technologies. Among its provisions are establishing 10,000 to 14,000 new public health information access points linked to an Internet-based HealthInterNetwork portal. Similarly, we are beginning to see kiosks and terminals devoted to linking citizens more closely to government information and communications, a service often labelled e-government, and illustrated by a system in India where once hard-to-get land records are now easily available in Karnataka state via self-serve computer and network equipped kiosks (Warschauer, 2003:175–77).

3.2 Demand-Driven

All three of these approaches to shared facilities depend for their ultimate success and sustainability on being demand-driven. Whether it is the situation reflected in the assertion that "widespread rollout of telecentres... can be achieved only by mobilizing private sector entrepreneurship and investment" (Wellenius, 2003) or that telecentres should get the same kind of public support (funding) given to libraries and other "public goods" kinds of community institutions (Gumucio Dagron, 2001a), telecentres, cybercafés and IAPs must serve the needs of a substantial portion of their host communities' population. As we have learned from the general demise of public extension systems around the world, even government support is not likely to be sustained, especially in difficult economic times, if there is not visible evidence of community patronage and benefits from these community enterprises.

In the remainder of this chapter we examine some of the factors that contribute to advancing a demand-driven approach, and, while the discussion focuses especially on telecentres, the issues will often be applicable to IAPs and cybercafés.

4. Aspects of Demand

4.1 Relevant Content

A demand-driven telecentre can form part of a solution to the need experienced by people in host communities for access to relevant and useful content. A telecentre may have low relevance if information is in unfamiliar or inappropriate language or dialect. Even where the mainstream language is English, there is evidence that this is not sufficient to attract people to ICT information resources. A study by the Children's Partnership in the United States looked at the extent to which currently available content met the needs of diverse communities. The study reported that the greatest barrier keeping low income people away from information portals was a lack of locally relevant information. The low income people needed such practical content as adult literacy programs, information on public benefits, easy-to-understand health encyclopaedias, consumer and credit information, and information related to employment and training (Warschauer, 2003).

The quality of information refers to its credibility, validity, trustworthiness, relevance, and usefulness. A 2002 multi-nation study by the International Institute for Communication and Development (IICD) suggests "easier access to globalized knowledge is fast turning us into consumers of distant and potentially irrelevant information" (Ballantyne, 2002). Local content, the report says, faces intense competition because big content initiatives tend to push their external content onto local communities. The study indicated the need to:

- Stimulate all kinds of local content expression for local application and use.
- Stimulate e-Content creation and communication for local and global use.
- Develop e-Content exchange and broadcast systems.
- Strengthen the "synthesis and adaptation" capacities of organizations working with both "global" and "local" content.

A case in India shows how the staff of a "village knowledge centre" dealt with the issues of local, relevance and language. The centres, established in Pondicherry on the southeast coast of India by the M. S. Swaminathan Research Foundation (MSSRF), demonstrated ingenuity, creativity and sensitivity in developing their information products. In one case, coastal villages were highly dependent on weather and tides information. Because many fishermen there were not literate, digital network information such as weather reports was downloaded and converted to audio by the village knowledge centre. The audio versions were then played on loudspeakers in the open air. In addition, project volunteers in the villages built their own information resources in the centre to complement the external databases, thereby providing local and localized information on agricultural, health and government programs for low-income people (Gumucio Dagron, 2001b). With the project staff, many locally useful databases were designed and developed, including, for example, a directory of general and crop insurance schemes; a list of about 130 schemes available as entitlements to rural families; a directory of hospitals and medical practitioners in Pondicherry, grouped according to their specializations; bus and train schedules covering Pondicherry and two nearby towns; and pest management information for the sugarcane crop (Harris, 2003).

4.2 Clustering and Networking

Creating appropriate localized content is very labour intensive, and without volunteer resources can be quite costly. The clustering of telecentres in some fashion can help support a localized information service. As the Pondicherry case suggests, MSSRF has made this arrangement, and the cost of producing local information is being spread over a number of telecentres in a relatively small area. Similarly in eastern Canada, community Internet access sites sometimes join together to make available French language material for local Francophone populations (Pfiester *et al.*, 2000).

Latin America's *Somos@Telecentros* Network (S@T) was one of the earliest significant regional community-based telecentres networks. After it had been in existence for two years, S@T published a study in which it listed the lessons it had learned. The very first on the list was: "No telecentre is an island". The study asserted: "If telecentres are to make their mission more effective, they need to organize themselves in overlapping national, regional and international networks" (Somos@Telecentros, 2003). *Somos@Telecentros* has done this, bringing together 350 telecentres in Latin America and the Caribbean area. This networking arrangement allows S@T to share resources and to provide access to resources more easily. Content appropriate for the region has been an important outcome.

4.3 Universities and the Demand Issue

One of the difficult parts of the content portal and web page approach is the continuous effort and investment needed to maintain them. The amount of voluntary effort to sustain a dynamic web resource with on-going interesting and useful content is substantial, and it has been suggested that only those with a significant financial stake or who are paid can make it work (Gurstein, 2001). To address this problem it may be necessary for universities to play a larger and more visible role in institutionalizing the generation and collection of development and local-oriented information resources (Colle and Roman, 2003).

The social role of the university has historically involved creating, storing and diffusing knowledge, a range of activities that partially parallels some telecentre operations. Nevertheless, few major programs link telecentres to universities as an institutionalized source of information. Some universities already have experience and commitments that are relevant to community development information and training through ICTs and telecentres. For example, universities have been involved in extension, a system designed to link researchers with potential users of their research. Furthermore, since the days of correspondence courses, universities have used a variety of media for distance learning, especially focusing on formal education at secondary school and college levels. However, beyond creating web pages, few universities have yet taken the step toward linking their knowledge resources to telecentres and to the potential of ICT for development.

Universities could play a significant role in addressing the priority actions suggested by the IICD study. For example, recognizing the "mixed quality" of relevant global and local content, there is the need to collate and screen the material to maximize its relevance and applicability (Ballantyne, 2003). Universities, like the network of provincial agricultural universities in China, typically have the intellectual resources and students who can work closely with telecentres and information portals to make the appropriate transformations (Colle and Liu, 2002). Universities are also in a unique position to address another factor related to demand—the training of community members in the use of information technologies. In addition, universities can be an important resource in undertaking community level research that is vital to making telecentres demand-driven—a subject that now takes us into the field.

4.4 Researching Demand

Harris (2003) describes an activity in East Malaysia that is essential to creating a demand-driven telecentre. Prior to the establishment of a telecentre in the small settlement of Bario in Sarawak (Borneo), the project collected data on the information needs of the community. The data reflected the type and amount of information members of the settlement would like to receive, what they were currently receiving, the type and amount of information they were sending, and the sources and channels used. The survey data revealed that "the community placed most importance on information relating to agricultural, medical and religious practices [with] information technology, job opportunities, government policies and family matters rated slightly less important" (Harris, 2003). In addition, using Participatory Action Research (PAR) methods, project leaders and the community were able to agree on a "prioritized set of information needs". This resulted in one person's action in assembling and documenting best practices for the production and treatment of Bario rice for which demand outstripped supply, thus not only responding to an information need but also providing a local information resource.

Research is important in identifying needs that can be factored into decision making about content made available through various ICTs at a telecentre. However there is another role that research should play in a demand-driven telecentre approach. Although there is a lot written and said about the potential of ICT for rural development, to this date there is only a meagre inventory of documented development outcomes resulting from telecentre initiatives. Research evidence of impact is at best sparse and anecdotal. Some authors are very critical about the scarcity of evaluative evidence about impact and the lack of assessment of local needs (Van Audenhove, 2000).

4.5 A Case of University Involvement

The Tamil Nadu University of Veterinary and Animal Sciences (TANUVAS), in co-operation with Cornell University, has created a network of rural telecentres in the state of Tamil Nadu, India, with the financial support of the International Development Research Centre of Canada. In this section, we present a brief summary of our research experience in this project, and some implications for demand-driven telecentres. During the summer of 2001, a TANUVAS-Cornell team, supported by staff from the Swaminathan Foundation (which founded the Pondicherry project mentioned earlier), conducted an information and communication needs assessment study. The research collected qualitative and quantitative data through a survey questionnaire and focus groups of local women and men in the three villages where telecentres had been established. Approximately 750 persons were interviewed.

The analysis of the focus group exercises shows differences in information uses and patterns depending on gender, age and occupation. Agriculture is the main economic activity of the villages studied. Farmers—men and women—require information on new seeds and products, fertilizers, market prices, and other agriculture-related issues. However, women especially demand information about their children's education and health, while the younger people are mainly interested in employment opportunities.

The research shows that there are cyclical changes in information needs during the annual calendar because village economic and social life revolves around agricultural seasons and local religious and cultural traditions. This situation has implications for the content provision activities facilitated by a telecentre. For example, men and women express an interest in employment opportunities during the months when there is not much activity in the fields. The high rates of illiteracy and low levels of formal education in the villages studied confirmed that content should be provided in Tamil.

4.6 Demand and Gender

The cultural barriers that hinder women's access to ICTs, and especially computers and the Internet, are more problematic and complex than simply making computers available in a library, telecentre or other public facility. Those barriers include literacy, education, language, cost, locality, the perceived role of women, and technophobia (Hafkin and Taggart, 2001). These are not inherent in the female condition nor are they barriers uniquely experienced by females. But they are barriers that exist widely and more severely for women particularly in Africa, Asia, and Latin America. Some of these obstacles are deeply embedded in cultural practices, such as denying school opportunities for girls.

In India there are hundreds of thousands of poor women attached to self-help groups (SHGs) that are involved in a wide array of micro-economic enterprises. Many have been mobilized by NGOs that have a commitment to improve the welfare of their constituents. For example, in Gujarat, the Self Employed Women's Association

(SEWA) has a membership of more than 200,000 women in some 790 villages. SEWA helps these women organize into groups or co-operatives so that they can co-operate to build stronger enterprises. The promotion of women's SHGs is seen as an effective means to empower poor women and enable them to participate in and drive their own development. SHGs are now recognized as a key transmission belt for development efforts by the state and the civil society. Development authorities prefer such village level collectives as institutional mechanisms because they are gender sensitive, participatory, and cost-effective grassroots organizations.

Cultural barriers prevent many of the women in these groups from benefiting from ICTs. TANUVAS, with UNESCO support, initiated a project in 2002 to broaden low income women's access to ICTs by having representatives of their SHGs trained in ICT use. This involved activities such as using the web to search for information, using e-mail, and working with self-learning and distance learning multimedia packages. An initial step was to identify the women's and the SHGs' needs and build the appropriate (responsive) information and other services into the existing telecentre operations.

5. Demand and the Telecentre Environment

There are aspects of the telecentre's environment that are likely either to drive demand for telecentre services or to drive people away. A variety of factors influence how readily the community will be motivationally and physically driven to go to the telecentre.

5.1 A Nice Place to Be

A very obvious point (though frequently ignored or overlooked) is the importance of making a telecentre a nice place to be. Telecentres can learn about this from some of the better cybercafés (Proenza, 2001). We studied one of Canada's community access facilities and found that separate times had to be scheduled for adults and young people because each was intimidated by the other. This made each group more comfortable. In another Canadian situation, we were told that just changing the name from "Community Access Program Site" to "Cybercafé" increased the visibility and use of the facility.

The BusyInternet (BI) telecentre in Accra (Ghana) takes the issue of atmosphere seriously. To attract people to the centre who might not otherwise be interested in technology, movies are shown at the centre on weekends. Another magnet is *Liquid*, the BI Accra restaurant and bar with its cool-blue bubble design. This is where the local cyber crowd hangs out to network and dream up ideas. The BI philosophy is that creating a social scene around technology will help spark an innovative technology culture, and it places equal importance on both social and financial return. For example, to raise awareness about national ICT policy, the telecentre hosts monthly debates and organizes lectures by experts. Low or no-cost Internet access is offered to those attending HIV/AIDS workshops and other socially oriented programs. Those who cannot afford the normal daytime prices of fee-based services can pay half-price at night (BusyInternet, 2003).

Do community members have problems in accessing the centre? Where is the telecentre located? It is clear that if the telecentre is far away from the usual community meeting points, it might hinder participation. In South Africa, the telecentre in the township of Mamelodi, in Pretoria, was originally located in the local library. Shortly after it was established, they decided to move to an independent location because the library location appeared to the community as an official or government site, which intimidated many who considered it to be for "intellectual people".

5.2 Participation

Building an atmosphere of community participation and ownership (though not necessarily in the literal sense) is important for the demand-driven formula. Participation has very practical value for telecentres. It comes in various forms including participants as telecentre users, participants as telecentre staff volunteers, and participants as telecentre advisory groups. Participant volunteers contribute to the daily supervision of the facilities—a potential personnel expense that many telecentres could not afford. Volunteers such as high school and college students, retired business people, schoolteachers, and senior citizens also provide telecentre clientele with personal models with whom they can identify and feel comfortable. In some places, women do not feel welcome in a telecentre because of the "maleness" of the environment and the accompanying intimidation. The presence of self-confident women volunteers helps overcome some of these obstacles.

An important result of our needs assessment in India was the creation of a local steering committee for each telecentre. These committees, formed by a diverse group of villagers (including people of both sexes, young and old) are in charge of monitoring the economic and social sustainability of the telecentres; they remain in close contact with personnel at TANUVAS. For example, these committees decide about new services by taking the pulse of village needs, and they administer existing resources and look for new ones (including looking for volunteers in the community interested in collaborating in telecentre activities). The steering committees act as local telecentre champions and agents of demand actualization. The challenge for telecentres is to move from largely spontaneous use and management of volunteers to the development of an explicit strategic plan for recruiting, training, and rewarding volunteers.

5.3 Awareness of the Information Society

Participation of community members in telecentre activities is problematic unless they are aware of what a telecentre is and understand its potential to improve their lives. Most villagers in India do not know either what a computer or the Internet can do for them or what they are. The people who are most in need of an information or communication service may not respond just to the service becoming available. In India, village elders and leaders traditionally act as the main source of information and communication. If a telecentre ignores this tradition, it may bring power clashes and conflicts. Social and economic divides may also hinder equal access to useful information services. Thus, awareness campaigns may need to deal with self-efficacy. Examples include women's self-assessment of their ability to take action, and women's perceptions of how significant people in their environment support or resist what the women do as a result of information gained.

Information and public relations campaigns promoting telecentres are part of the solution but the parallel challenge lies in the appropriate development of services (Clark, 2001). Telecentres can systematically assess community information needs and the communication needs of various local organizations, and be creative and entrepreneurial in dealing with these needs. It is this more comprehensive community service approach to the Information Society that helps telecentres become more firmly woven into the fabric of the community and puts them on the road to being demand-driven and sustainable.

5.4 Intermediaries

Richard Heeks (1999) of the Institute for Development Policy and Management at in the University of Manchester suggests that intermediaries are organizations or individuals "who own ICTs and who can act as gatekeepers between cyberspace and the organic, informal information systems of those on the wrong side of the digital divide". Heeks suggests that good intermediaries bring more to the process than connection to information and communication data and hardware. Moreover, whilst ICTs can deliver potentially valuable information to end-users, they may not possess the ability to act on it. Market information is useless if it is not possible to transport goods, and medical advice is meaningless if there is no money to purchase medicines. Community organizations and institutions can create demand for telecentre services. Schools, health centres, agricultural extension agents and input suppliers, community leaders, and co-operatives should be partners with telecentres in identifying what communities need in order to be able to act on information. Telecentre managers must reach out to community groups and demonstrate how telecentre resources apply to development activities. Agricultural extension, community health workers, school-teachers and government officials need to re-examine how information technology can contribute to their efforts.

6. Conclusion

Macro data linking telecentres to improvements in a nation's profile in the United Nations Development Program's Human Development Report do not exist. However, individual stories of telecentre impact are quite persuasive. We expect to see new ways in which demand can drive telecentres and perhaps enable them to subsidize some of the development-oriented services that some people cannot afford. For example, the spread of E-Commerce presents an opportunity to use telecentres as a place where rural people could order goods on the Internet, and later pick them up and pay for them, for which the telecentre could charge a transaction fee. Similarly, expatriated workers sending remittances to families in developing countries serves as a source of telecentre income (Robinson, 2001). Despite pessimism about the cost effectiveness of telecentre investments (Kenny 2003; Gómez and Casadiego, 2002; Mardle, 2003), it is evident that nations that ignore their potential risk falling behind in rural development. The relevance and sustainability of telecentres depends on how they can connect their services with the demands of their communities, and the extent to which populations can be convinced that their needs and desires can be translated into telecentre services.

References

Andresen, A. (1995) *Marketing Social Change*. San Francisco: Jossey-Bass Publishers.

Ballantyne, P. (2002) *Collecting and Propagating Local Development Content*, Research Report No. 7. The Hague: International Institute for Communication and Development.

Batchelor, S. (2002) *Using ICTs to Generate Development Content*, Research Report No. 10. The Hague: International Institute for Communication and Development.

Bihari, G. and Jókay, C. (1999) *Telecottages in Hungary: The Experience and the Opportunities*. Budapest: I.G.E. Ltd..

BusyInternet Accra: ICT-Enabled Services Case Studies Series. (2003) www.bridges.org/iicd_casestudies/busy_internet/

Clark, J. (2001) Promoting Participation in Telecentres, *The Journal of Development Communication*, **12**, 2, ip.cals.cornell.edu/commdev/documents/jdc-clark.doc.

Colle, R. (2003) Threads of Development Communication, in: Servaes, J. (ed.) *Approaches to Development Communication*. Paris: UNESCO. [Also on CD].

Colle, R. and Liu, Y. (2002) ICT Capacity-building for Development and Poverty Alleviation; The Role of Agricultural Universities in China, in: Mei, F. (ed.) *Asian Agricultural Information Technology and Management: Proceedings of the Third Asian Conference for Information Technology in Agriculture*, Beijing, October 26–28

Colle, R. and Roman, R. (2003) Challenges in the Telecentre Movement, in: Marshall, S., Taylor, W. and Yu, X (eds.) *Closing the Digital Divide: Transforming Regional Economics and Communities with Information Technology*. Westport: Praeger.

Gáspár, M. (2001) Telehouses in Hungary, *The Journal of Development Communication*, **12**, 2, ip.cals.cornell.edu/commdev/documents/jdc-gaspar.doc.

Gómez, R. and Casadiego, B. (2002) *Letter to Aunt Ofelia: Seven Proposals for Human Development Using New Information and Communication Technologies*. Ottawa: International Development Research Center.

Dagron, G. A. (2001a) Prometheus Riding a Cadillac? Telecentres as the Promised Flame of Knowledge, *The Journal of Development Communication*, **12**, 2, ip.cals.cornell.edu/commdev/documents/jdc-dagron.doc.

Dagron, G. A. (2001b) Making *Waves: Stories of Participatory Communication for Social Change*. New York: Rockefeller Foundation.

Gurstein, M. (2001) Commentary: The Global Development Gateway and the Central Dilemma of E-Commerce, in: The Communication Initiative, www.comminit.org.

Hafkin, N. and Taggart, N. (2001) *Gender, Information Technology, and Developing Countries: An Analytic Study*. Washington D.C.: Office of Women in Development, U.S. Agency for International Development; Abridged version available in: Ensuring Women's Ability to Take Advantage of Information Technology Opportunities, *The Journal of Development Communication*, **12**, 2.

Harris, R. (2003) Information and Communication Technologies for Rural Development in Asia, in: Marshall, S., Taylor, W. and Yu, X (eds.) *Closing the Digital Divide: Transforming Regional Economics and Communities with Information Technology*. Westport: Praeger.

Heeks, R. (1999) *Information and Communication Technologies, Poverty and Development*, Working Paper Series of the Institute for Development Policy and Management, University of Manchester, www.man.ac.uk/idpm.

Herz, J. C. (2002) The Bandwidth Capital of the World, *Wired*, **10**, 08 www.wired.com/wired/archive/10.08/korea.html.

Kenny, C. (2002) Information and Communication Technologies for Direct Poverty Alleviation: Costs and Benefits, *Development Policy Review*, **20**, 2, 141–157.

Mardle, E. (2003) Telecentres: How Did We Lose the Plot, Development Gateway. www.developmentgateway.org/node/133831/sdm/docview?docid=440944

Pfiester, A., Roman, R. and Colle, R. (2000) The Role of Participation in Telecentre Initiatives, *The Journal of Development Communication*, **11**, 2, www.aidcom.com.

Proenza, F. (2001) Telecentre Sustainability: Myths and Opportunities, *The Journal of Development Communication*, **12**, 2, ip.cals.cornell.edu/commdev/documents/jdc-proenza.doc.

Robinson, S. (2001) Rethinking Telecentres: Knowledge Demands, Marginal Markets, Microbanks, and Remittances, *On the Internet*, thinkcycle.media.mit.edu/thinkcycle/main/assisting_microbanks/challenge___telecentre _transaction_support/0401robinson.pdf

Rodriguez Oliveros, M. (2002) The National e-Mexico System: Factors of Success and Failure, unpublished paper, Ithaca: Cornell University.

Roman, R. and Blattman, C. (2000) Research for Telecentre Development: Obstacles and Opportunities, *The Journal of Development Communication*, **10**, 2, www.aidcom.com.

Singhal, A. and Rogers, E. (1999) *Entertainment-Education: A Communication Strategy for Social Change*. Mahwah: Lawrence Erlbaum Associates.

Somos@Telecentros (2003) The Story So Far and Lessons Learned, www.tele-centros.org/english/new/index.html

Van Audenhove, L (2000) Information and Communication Technology Policy in Africa: A Critical Analysis of Rhetoric and Practice, in: Avgerou, C. and Walsham, G. (eds.) *Information Technology in Context: Studies from the Perspective of Developing Countries*. Ashgate, Burlington.

Warschauer, M. (2003) *Technology and Social Inclusion*. Cambridge: The MIT Press.

Wellenius, B. (2003) Sustainable Telecentres: A Guide for Government Policy, *Public Policy for the Private Sector, Note Number 25*, Washington D.C.: The World Bank Group.

About the Authors

Royal Colle has been on the faculty of Cornell University (USA) for almost 40 years. Although a professor emeritus, he continues to teach there and work on issues related to new information technology and development. Colle has served on a variety of projects in health, agriculture, nutrition and family planning for international organizations including the World Bank, the World Health Organization, UNFPA, ITU, FAO and USAID. In recent years he has concentrated on telecentre projects in China, India and several African nations.

Raul Roman has been the lead evaluation specialist on Cornell University's project called The Essential Electronic Agricultural Library, an initiative designed to provide an "instant" digital agricultural research library for institutions in developing nations. Raul's Ph.D. dissertation focused on information needs analysis in India villages. He and Colle have jointly authored a Handbook for Telecentre Staffs, which was supported by the International Telecommunications Union and the Food and Agriculture Organization. It is available online at: ifp.cals.cornell.edu/commdev/handbook.cfm.

8. Regional Information Centres in Azerbaijan—
A Preliminary Evaluation

Michele Cocchiglia

International ICT Expert
E-Government for Development Technical Unit of the Government of Italy
Michele.Cocchiglia-alumni@lse.ac.uk

Abstract

Azerbaijan's ICT sector is one of the most rapidly developing in the country, and is receiving growing attention from a number of national and international institutions. In line with this trend, and in order to ensure a more widespread and equitable access to information and communication services, the Government of Azerbaijan and UNDP are currently establishing a number of Regional Information Centres (RICs) throughout the country. Regional Information Centres are shared facilities providing access to the Internet and ICT-enabled services, with the aim of promoting local and regional development. They are, however, a relatively new institution in Azerbaijan, and despite their potential to advance the country's overall development, their exact role as a development tool is in need of refinement and operationalization. This preliminary evaluation was an attempt to further investigate both the opportunities offered by RICs in Azerbaijan, and the challenges these facilities will face in the future, in light of the global experience gained with similar initiatives. The major issues of concern for the study were related to sustainability, and to the potential developmental impact of the centres on the recipient communities.

The views expressed in this article are those of the author, and should not be attributed to the E-Government for Development Technical Unit.

1. Introduction

1.1 ICTs for Development in Azerbaijan

Information and Communication Technologies (ICTs) are playing an increasingly important role in Azerbaijan's social, economic and political development. At the present time, the ICT sector is one of the most rapidly developing in the country, and is receiving growing attention from a number of national and international institutions.

Global Experience clearly indicates that Information and Communication Technologies are an effective tool for assisting a country's successful development. This is considered particularly relevant for countries, such as Azerbaijan, undergoing a socio-economic transition, and aiming at a successful integration into the world community and information society.

In line with this vision, the Government of Azerbaijan and UNDP have recently developed a "National Information and Communication Technology Strategy" (NICTS) for 2003–12, which was adopted by the President on 17 February 2003. The Strategy identifies the country's key goals and objectives in this specific area, and sets priorities and main activity directions to guide the implementation of ICT for development projects. Azerbaijan's NICTS is expected to be one of the most important components of the country's overall economic, social and political progress.

One of the main directions of the National ICT Strategy is the elimination of the "digital divide" within the country. Today, there are still considerable differences between Baku and the regions of Azerbaijan in terms of spread and access to ICTs, and a number of factors—ranging from low incomes per capita to poor infrastructure development[17]—are currently hindering a more extensive use and diffusion of ICT applications.

1.2 Regional Information Centres and Rural Access

Considering the nature and diffusion of ICTs, the Government of Azerbaijan and UNDP are currently establishing a number of Regional Information Centres (RICs) throughout the country, so as to ensure a more widespread and equitable access to information and communication services.

Regional Information Centres are shared facilities providing access to the Internet and ICT-enabled services, with the aim of promoting local and regional development. Their potential for fighting against the marginalization of rural or otherwise disadvantaged areas, by fostering the dissemination and creation of relevant information and knowledge, is widely recognized (Fuchs, 1998; Latchem and Walzer, 2001; Etta and Parvyn-Wamahiu, 2003).

---

[17] According to a recent survey conducted in August and September 2003 in seven regions of the country—Ganja, Mingechevir, Ali-Bayramli, Sheky, Guba, Lenkoran and Barda—only 0.15 Internet-hosts come to each 1000 people, as compared to 4 to each 1000 in Baku (UNDP, NHDR Azerbaijan 2003, State Statistics Committee).

The rationale for the establishment of these facilities in the Azerbaijani context is twofold. On the one hand, Regional Information Centres are seen as an effective way to provide valuable information services to a larger segment of the population, and to increase the general level of awareness of information and communication technologies. On the other hand, RICs are expected to have a positive social and economic impact on the communities they serve, through the development of new skills and capacities. Qualification level and employability of the local population, are the major focus in pursuit of this latter goal.

1.3 Objectives of the Study

Regional Information Centres are a relatively new institution in Azerbaijan, and despite their potential to advance the country's overall development, their exact role as a development tool is in need of refinement and operationalization.

The aim of this preliminary evaluation was to further investigate both the opportunities offered by RICs in Azerbaijan, and the challenges these facilities will face in the future, in light of the global experience gained with similar initiatives. The major issues of concern for the study were related to sustainability, and to the potential developmental impact of the centres on the recipient communities. Seven[18] of the ten Regional Information Centres recently established in Ali-Bayramli, Guba, Imishly, Khanlar, Kyurdamir, Lenkoran, Shamakhy, Sheky, Sumqayit and Yevlakh, were visited for this purpose in December 2003 (see Appendix).

Finally, considering the general lack of information on RICs and similar initiatives in CIS countries, the study was also seen as a step forward towards the identification of regional requirements and specificities of these facilities.

2. The Global "Telecentre" Experience

2.1 Towards a Definition

Regional Information Centres are part of a larger movement, which is usually referred to as the "Telecentre" movement. Telecentres are recognized and called by a large number of different names, and are receiving increasing attention and support from various institutions, including international organizations, national governments, telecommunications operators and service providers.

In general, Telecentres are established for development purposes. They have as their aim the social and economic development of remote and otherwise disadvantaged communities, which is usually achieved through the provision of ICT-enabled information and communication services. Pilot Telecentre projects are currently being

---

[18] Regional Information Centres forming the subject of investigation were the ones established in Ali-Bayramli, Khanlar, Kyurdamir, Shamakhy, Sheky, Sumqayit and Yevlakh.

implemented in many developed and developing countries across the globe, with significant differences in terms of size, services provided, ownership and operating models, and according to their urban or rural location. As a result, no single clear definition of the concept exists, as the final form and function of these facilities is still being created, and is shaped by the specific context of implementation.

Despite these important differences, however, common Telecentre experiences and lessons are emerging from the field, highlighting critical issues, and providing recommendations to guide the practical implementation of these projects.

2.2 Key Issues and Trends

To date, the debate about Telecentres in developing countries has mostly focused on their financial sustainability. It has only recently become clear that, in order for Telecentres to be successful, sustainability has to be addressed taking into account its many-sided nature. According to this perspective, a solution for Telecentre sustainability is likely to emerge only by looking further than the simple availability of financial resources, and thinking about the conditions needed for Telecentres to be sustainable from different perspectives (Mayanja, 2002; Stoll, 2003). This is considered particularly relevant if the Telecentre is used as a tool for achieving community development, and does not only focus on the provision of ICT equipment and services.

More specifically, this has suggested the need to widen the concept, and to consider the social, political, and technical dimensions of sustainability as equally important elements. Most recently, a number of factors have been associated with sustainability, and include such aspects as the overall operating environment (i.e., socio-political context, technological environment, etc.), ownership and management styles of the Telecentres, community involvement, and relevance of services and content (Roman and Colle, 2002; Etta and Parvyn-Wamahiu, 2003).

This is not to say that financial sustainability is not important. Indeed, although it is only one of several dimensions, it remains the most questioned and possibly the most problematic. To date, Telecentre initiatives in developing countries have mostly been financed and supported by external donors, and often struggled to become financially independent. Private sector involvement has been rather limited so far, and usually restricted to donations and contributions. A number of experts and practitioners have stressed the importance of involving the private sector, assuming that if not operating as a commercial and profitable organization, Telecentres will simply encourage incompetence and dependency, eventually leading to losses and failure (Best and Maclay, 2001; Proenza, 2001).

On the other hand, it has been argued that if Telecentres have to serve and match the needs of communities, they should be somewhat considered as a "public good", worth supporting regardless of commercial viability, for the benefit of current and future generations. The evident weakness of this position, however, is that the conceptual validity of this argument does not necessarily ensure or lead to financial sustainability, nor to the achievement of positive developmental results.

Regardless of the operating and conceptual model adopted, however, overall economic viability still remains an important goal. An increasingly accepted idea is that in order to generate income and be considered successful, Telecentres should be demand-driven, and that this demand should be reflected in the community's willingness to pay for some of the services provided (Fuchs, 1998; Roman and Colle, 2002). Although this is a reasonable expectation, it is directly linked to other complex issues (e.g., relevance, accessibility, etc.) and is, therefore, not easy to achieve. One of the biggest challenges Telecentres face, for example, is the provision of relevant information and services to the recipient communities. These should be appropriate and specific, and preferably developed with local partners who can contribute to this process.

Overall, a generally accepted indicator of a Telecentre's success is the extent to which it becomes part of the community it serves (Fuchs, 1998; Roman and Colle, 2002; Mayanja, 2002). According to this perspective, people in the communities should feel empowered by the center, and actively involved in meeting the challenge of sustainability.

Other important factors affecting the performance of a Telecentre, are operative at the micro and macro socio-political levels. These include such aspects as the overall national policy environment, and the local arrangements for the management and control of facilities. At the Telecentre level, for example, sound management, as well as genuine community support, appear to be critical to their success (Etta and Parvyn-Wamahiu, 2003). Finally, a number of factors, ranging from affordability of the services provided to literacy level of Telecentre users, are widely acknowledged as potential impediments to use, and in need of careful consideration and monitoring (Harris, 1999; Roman and Colle, 2002; Etta and Parvyn-Wamahiu, 2003).

3. Assessment of Regional Information Centres in Azerbaijan

3.1 Equipment and Services Provided

At the time of the study, most of the Regional Information Centres were still in their early days of operation. Nevertheless, they were generally found to be well equipped and furnished. All the centres were being housed in buildings provided, and often refurbished, by the local government (i.e., executive authorities and municipalities). Despite their location in premises originally not designed for this type of use, the centres were providing sufficient space and had been adequately arranged for the new intended purpose. Most of the centres, however, were still not sufficiently heated.

A major issue of concern in this regard appeared to be responsibility over the maintenance of premises. Only some of the local government representatives, for instance, had agreed to provide the facilities with electricity, and to ensure the full coverage of electricity-related costs. This situation had serious consequences on the functioning of some of the facilities. In Khanlar, for example, the center had not been provided with electricity, and in spite of the fact that equipment and furniture had been delivered, the center could not become operational. A similar situation was found in Yevlakh, where electricity was available for only one hour per day during the working hours of the center. As a result, only theoretical lessons had been delivered for a few hours per week.

One of the most positive elements, on the other hand, was the very good quality of the furniture and equipment available. On average, 13–14 computers were found in each of the centres, along with one server, a scanner, a printer, and uninterrupted power supply (UPS) equipment (Table 8.1). The majority of computers were using one of the latest versions of Microsoft© Windows operating system, and running applications of common use, such as word processors, spreadsheets and presentation software in English language.

None of the centres, however, had a generator; taking into account the overall quality and reliability of power supply in the regions, this would have often proved extremely useful for running (part of) the electrical equipment, and ensuring an uninterrupted provision of services. Furthermore, although some of the centres had established a temporary dial-up connection, all of them were still lacking an adequate and reliable connectivity. In spite of the fact that a deal had been made with one of the private ISPs offering Internet services in the regions, Internet traffic over the public (state-owned) lines had not been enabled yet by the Ministry of Telecommunications. Services made available to the communities were therefore limited to training in computer and Internet basics, and office applications.

Table 8.1
Equipment and Services Provided by the Regional Information Centres

Region	Server(s)	PCs	Printer (s)	CD-RW	Internet Connectivity		UPS	ICT Training
					Established	Planned		
Ali-Bayramli	1	17	1	1	Dial-up	128 Kbps	●	●
Guba	1	15	1	1	–	128 Kbps	●	–
Imishly	1	11	1	1	–	128 Kbps	●	–
Khanlar	1	8	1	1	–	128 Kbps	●	–
Kyurdamir	1	13	1	1	Dial-up	64 Kbps	●	●
Lenkoran	1	12	1	1	–	128 Kbps	●	–
Shamakhy	1	12	1	1	Dial-up	128 Kbps	●	–
Sheky	1	14	1	1	–	128 Kbps	●	●
Sumqayit	1	20	1	1	Dial-up	128 Kbps	●	●
Yevlakh	1	15	1	1	–	128 Kbps	●	●

All the training courses provided had been developed in Azeri, based on the standards of the European Computer Driving Licence (ECDL). A few copies of an educational software in Russian, specifically designed for users with minimum computer experience, were also found in each of the RICs. Services were being provided at no cost, and would have been free of charge for the first year of operation at least. In addition to the existing services, further training, including foreign language courses to be developed with external partners, were also on the agenda of the project. Overall, services offered by the newly established facilities appeared to be greatly appreciated by the local communities. In Ali-Bayramli, discussions with the manager of the Regional Information Center revealed that more than 160 applications for computer training courses had been received, for a limited number of places available. Furthermore, and perhaps most importantly at this stage, support provided by the staff appeared to be effective and professionally delivered.

3.2 Management and Ownership of Facilities

Each Regional Information Center had a total of four staff members, which were responsible for day-to-day operations. Among them, three were providing technical assistance and computer training, and one had been identified as the person responsible for the management of facilities. Staff had been hired among the local population on the basis of their technical and professional background, and generally appeared to be sufficiently skilled and highly motivated. Their exact duties and responsibilities, however, were still not adequately explicit, especially with regard to management and administrative procedures. As a result, the project implementers were still in full control of budget and activities.

Although this was partially due to the early stage of implementation of the centres, clearly defined roles for the staff members appeared to be urgently required. Critical aspects to be defined included, for example, responsibilities over the operating costs, equipment replacement and maintenance, as well as clear and functional financial procedures.

If Regional Information Centres are to be effective in serving the needs of their users, they will require the power and flexibility to operate independently, facing specific problems and demands with the resources they have at their disposal. As it is clear, definition of duties and procedures will become even more critical if and when the centres will stop running their operations exclusively relying on donor funds. Furthermore, responsible and independent management is also seen as instrumental to financial sustainability, on account of the weaker motivation generally shown by donor-driven projects to become self-sustaining (Mayanja, 2002).

Another aspect which certainly deserves further attention is the extent and nature of community involvement. As stated earlier, this is considered critical to the success of the centres, especially with regard to their development component. At the time of the study, community involvement appeared to be fairly weak, and effective demand for information and services still poorly investigated. Despite the fact that the services provided had been positively received by the local communities, the major focus ap-

peared to be on the provision of educational services, regardless of the potential existence of other (unexpressed) needs. To a certain extent, this is symptomatic of the
type of approach adopted for the development of facilities, in a country with a relatively short history of community development, and an extremely complex socio-
political environment.

In many facilities, community-based organizations or committees are used as a
tool to empower the target communities. These are usually given responsibilities such
as representing different community interests and groups, and "overseeing" the center's activities and operations. Such organizations might be difficult to establish in the
Azerbaijani context, and local development may be more easily promoted in a different manner—observing and responding to, for example, the changing needs expressed
by actual users of facilities. If similar organizations to the ones suggested are to be
created, however, transparent criteria for member selection will need to be adopted,
so as to genuinely voice and articulate information and communication needs of the
recipient communities.

3.3 Financial Sustainability

Financial sustainability of the centres was still difficult to assess at the time of the
study. A first estimate of the operating costs is provided below, based on the information available (Table 8.2). As it is often the case, monthly expenditures of the RICs
will largely depend on the type and speed of Internet connection established. Internet
usage intensity[19], and to a smaller extent, power consumption, will also affect the
overall budget requirements.

Furthermore, along with operational costs, additional and unexpected expenditures, arising for instance from technical maintenance and upgrade of equipment, are
also to be considered, and difficult to predict at this stage. A significant portion of
these costs will depend on such aspects as the frequency and gravity of technical
problems, and on the staff ability to deal with them independently (i.e., without relying on external sources).

Table 8.2
Estimated Average Monthly Expenditures, According to Internet Connection

Estimated average monthly expenditures	Staff (4 persons)	Electricity*	Internet Connectivity*	Consumption Material*	Total*
	50$ x 4=200$	125$	590$ – (64 Kbps)	20$	935 $
			950$ – (128 Kbps)		1,295 $

Estimated value

----

[19]   The monthly cost of a leased line is calculated according to the percentage of bandwidth used.

As mentioned above, the most significant hurdle for financial sustainability is the high cost of Internet access. This represents by far the major expenditure, accounting for up to 70% of the estimated total. In some measure, this is the result of insufficiently promoted institutional and structural reforms. Despite the fact that privatization has commenced in Azerbaijan, and Internet access prices have shown an encouraging decreasing trend[20], state-owned telecommunications service providers still enjoy a monopolistic and unfair competition environment. Formal and informal interferences of these providers with the business of private operators are frequent in the country, and represent a major obstacle to a greater development of the ICT sector. On account of this absence of competition, communication lines remain poorly developed in the regions, and tariffs significantly high[21]. Difficulties experienced by the Regional Information Centres with the establishment of Internet connection, are also a result of this unfavorable environment.

Another major challenge RICs will face in their attempt to become self-sustaining is the low purchasing power of the local population. Low average salaries and a high unemployment rate are a serious problem in Azerbaijan, and a decision to charge users for the services provided might result in little and insufficient demand, even at below-market rates.

Evidence suggests that successful telecentres usually become self-sustaining after 3 or 4 years of operations, while relying on external funds to cover part of their costs during their first few years of existence (Fuchs, 1998; Jensen, 2001). It is reasonable to expect that Regional Information Centres will be no exception in this respect, and that substantial efforts and time will be required to achieve this goal. What can be said at this early stage is that the overall financial performance of the centres will largely depend on their ability to maintain a good control over expenditures, to attract subsidies, donations and grants, and to identify profitable services, ensuring commitment and sensitivity to local needs.

3.4 Relevance and Accessibility

As mentioned above, although some efforts had been made to address and match local community needs, these were still at the incipient stages, and in need of deeper consideration and analysis. The overall impression was that a strong emphasis was placed on the provision of educational services, while potentially unexpressed needs and expectations of the target communities were still taken into little consideration. If Regional Information Centres are expected to act as a development agent, however, appropriate content and relevant services will need to be developed, and constantly adapted to the communitie's changing needs. As previously underscored, simple provision of infrastructure is unlikely to stimulate development, unless "smart" services

---

[20] According to the Azerbaijan Development Gateway, Internet access prices in the country have decreased from $2 per hour in 1999 to $0.50 in 2001.

[21] Among the 12 CIS countries, for example, Azerbaijan has the second-highest prices for international calls (Azerbaijan Human Development Report 2003).

are tailored around it. Services and content are generally seen as complementary to infrastructure, and instrumental to the achievement of specific development goals (Roman and Colle, 2002; Mayanja, 2002). Moreover, without consideration of this aspect, attempts to encourage a greater use of the centres may be met with limited success, with undesirable consequences on their financial performance.

Training courses, such as the ones provided by the Regional Information Centres, are indeed relevant for promoting local development, although they only represent one of the potential services these facilities can deliver, and only target a specific development goal. Furthermore, relevance of services is generally considered a context-specific matter. Services should be relevant for the targeted community, where "community" refers to a specific group of individuals, whose interests and needs are to be carefully identified and monitored.

Equally critical is the nature and level of access to the services provided. As mentioned above, this might be influenced by such aspects as the level of English and computer literacy, location of the facilities, and by social and cultural factors (Etta and Parvyn-Wamahiu, 2003). As previously highlighted, affordability is likely to be one of the major impediments to use, on account of the low levels of disposable income among the local population.

Another issue to be considered is the general lack of Azeri content available on the Internet, suggesting the need to create, "repackage" and translate existing information and content. The relatively high literacy rate of the local population, on the other hand, represents one of the most encouraging elements in this respect, giving a chance to develop more sophisticated and advanced services and applications.

4. Follow-up and Future Directions

4.1 Promoting Successful Partnerships

As is clear from the foregoing, in spite of some common elements, Regional Information Centres are likely to face different problems, and to find different solutions to these problems according to their specific features and to local conditions. Overall, the ability to promote strategic partnerships, and develop valuable services for the target communities is critical at this stage. For partnerships to be considered successful, however, two elements will ideally need to be preserved. First is independence from political interference, which has often proved a threat to community development (Proenza, 2001). Second is the ability to make decisions at the local level, for a number of reasons which have previously been discussed. Successful partnerships are also essential from a financial perspective. As evidence suggests, without the involvement of, and interaction with, external organizations, Regional Information Centres are unlikely to develop and retain a critical mass of users.

Grassroots NGOs, for instance, have proved an excellent vehicle for reaching the local communities. There is evidence to suggest that joint action for a common goal offers the potential for promoting development over and above the direct benefits generally associated with ICTs. Although co-operation and trust among different partners might be difficult to achieve at first, as the centres expand and evolve, they could develop a mutual understanding of their respective approach, and elaborate effective ways of working together on a joint purpose. Micro enterprise and small businesses also represent a possible partner to be involved in the development of Regional Information Centres, especially in some of the regions. These are potential users and promoters of services such as software training, marketing, and consultancies. In several countries, for example, small businesses and local entrepreneurs have used similar facilities to design local websites, advertise and promote their activities or develop basic business plans, while enriching the centres with valuable skills, knowledge, and through content creation (Latchem and Walzer, 2001; Etta and Parvyn-Wamahiu, 2003).

4.2 Developing Valuable Services

Integration of facilities into the local communities and development of valuable services are generally acknowledged as a required condition for the success of the centres, and are clearly relevant even from a financial perspective. Simple provision of common services (i.e., email access, Internet browsing, fax, photocopying, etc.), however, have often proved insufficient to generate adequate incomes, and the need to develop "smarter" solutions and services has been frequently pointed out (Mayanja, 2002; Etta and Parvyn-Wamahiu, 2003). Yet again, financial performance is normally influenced by local conditions, and in spite of the fact that a few models have been successfully developed in some specific contexts, none of these is likely to be replicable on a larger scale.

In some Eastern European countries, for example, a successful financial model has been the contracting out of facilities and services to international NGOs and other development agencies (Latchem and Walzer, 2001). The success of this model relies on the multiple benefits generated for the different actors involved, in terms of additional revenues for the centres, cost-efficiency gains for the contracting organizations, and delivery of appropriate services for the recipient communities. In other countries, similar facilities have covered part of their operating costs by providing, for example, advice to their users on how to apply for national and international grants, by selling space for small businesses to advertise at a modest price, or through the delivery of employment, tourism, and similar services (Latchem and Walzer, 2001; Etta and Parvyn-Wamahiu, 2003).

Government information and e-governance applications could also find their place in the future of Regional Information Centres. At present, e-governance in Azerbaijan is still in its infancy, and limited to a partial automation of the public administration and an initial presence of government institution on the Internet. Governmental websites in Azerbaijan do not allow yet real interaction with the citizens, and most of the documentation made available to the general public is still delivered in hard copy, mostly in the capital city. Hence, RICs have the potential to contribute to the filling of this gap, by providing information on the government's social, educational and other programs, and creating a stronger link between citizens and government institutions. Information delivery might be initially performed in a non-digital format, and possibly evolve into effective online interaction with government institutions.

Once again, however, regardless of the services and information delivered, effective demand and use remain the keys for success. Yet demand sometimes needs to be stimulated, especially in the beginning, and communities informed about the facilities and their potential benefits. Regional Information Centres "awareness days" or similar campaigns, aimed at educating the target groups on the services offered, could be an effective tool for community mobilization, and an important first step towards the achievement of a critical mass of users.

4.3 Creating a "Network" of Regional Information Centres

Another issue which is worth mentioning, especially considering the high number of facilities established in the country, is the creation of a network of Regional Information Centres. There is some empirical evidence indicating that "standalone" centres hardly are very successful. Integration of these facilities is increasingly pointed out as a required condition for their successful management (Roman and Colle, 2002; Stoll, 2003). This organization is generally suggested for several reasons. First, it increases effectiveness of the centres by allowing them to share insights, experience and "best practices", both through online and "face to face" encounters. Second, it offers a chance to share and more easily access resources. Third, a network of Regional Information Centres is likely to have a higher contractual power, and be more efficient in public policy debates. Being part of a local, national, if not regional network, for example, could allow the centres to achieve connectivity at lower costs, through block deals between the network and governmental institutions. Finally, a network of Regional Information Centres under a single "management", name or branding, might also carry the advantage of increased recognition.

5. Conclusions and Recommendations

A preliminary assessment of Regional Information Centres suggests that along with the existing achievements accomplished during their initial phase of implementation, a number of challenges will hinder their way to sustainability and effective development. The point is made that, ideally, RICs should serve as a community space. They should facilitate the diffusion, creation and exchange of information of general and specific interest, and provide appropriate services, ensuring sensitivity and commitment to local needs. In order to achieve this complex goal, several issues will need to be addressed. The following broad recommendations can be drawn from the foregoing, and act as a starting point:

- Regional Information Centres should develop a clear strategy to reach their objectives. Middle-term goals should be defined, so as to guide the implementation of activities, and facilitate future monitoring and evaluation;

- Effective needs and demand among the local communities should be further investigated and assessed, and valuable services and content developed accordingly;

- Potential partners for provision of such services and content should be identified, and effective operating and financial models attempted;

- Clear roles and duties should be defined for the centres' staff members, and functional operational procedures established;

- Linkages and dynamics between project implementers, local government representatives, and staff and community members should be further assessed and monitored;

- A more transparent policy and regulatory environment, leading to effective market competition, should be promoted;

- A network of Regional Information Centres could also be developed, so as to benefit from the advantages carried by this type of organizational structure.

Finally, as the centres evolve over time, additional elements, which were difficult to assess at this early stage, will require careful analysis and constant observation. These include such aspects as:

- The type of use (i.e., nature and diffusion) and users (i.e., gender, age distribution, etc.) of Regional Information Centres;

- The level of users satisfaction and the existence of potential impediments to use;

- The overall technological environment (i.e., frequency and gravity of technical problems, reliability of Internet connectivity, power supply, etc.);

- Staff members' skills and performance, and their ability to overcome common problems independently;

- The overall social and economic impact of Regional Information Centres on the recipient communities.

References

Azerbaijan Development Gateway (2001) *E-readiness Assessment Report—2001*, Baku, Azerbaijan.

Best, M. and Maclay, C. (2001) Community Internet Access in Rural Areas: Solving the Economic Sustainability Puzzle, in: Kirkman, G., Cornelius, P. K., Sachs, J. D. and Schwab, K. (Eds.) *The Global Information Technology Report 2001–2002: Readiness for the Networked World*, Oxford University Press.

Colle, R. (1998) The Communication Shop: A Model for Private and Public Sector Collaboration— Sustainable Tool, Paper Presented at the Don Snowden Program Conference: *Partnerships and Participation in Telecommunications for Rural Development: Exploring What Works and Why*, Guelph, Ontario, Canada.

Etta, F. and Parvyn-Wamahiu, S. (Eds.) (2003) *Information and Communication Technologies for Development in Africa: The Experience with Community Telecentres*. International Development Research Centre (IDRC), Ottawa.

Fuchs, R. P. (1998) Little Engines that Did—Case Histories from the Global Telecenter Movement, International Development Research Centre (IDRC), Ottawa, web.idrc.ca/en/ev-10630-201-1-DO_TOPIC.html.

Harris, R. (1999) Evaluating Telecenters within National Policies for ICTs in Developing Countries, in: Gòmez, R. and Hunt, P. (Eds.) *Telecentre Evaluation: A Global Perspective—Report of an International Meeting on Telecentre Evaluation*, International Development Research Centre (IDRC).

Jensen, M. (2001) *The Community Telecentre Cookbook For Africa: Recipes For Self-Sustainability— How to Establish a Multi-purpose Community Telecentre in Africa*. United Nations Educational Scientific and Cultural Organization (UNESCO), Paris, France.

Latchem, C. and Walzer, D. (Eds.) (2001) *Telecentres: Case Studies and Key Issues,* Commonwealth of Learning Publication, Vancouver, BC.

Mayanja, M. (2002) *The African Community Telecentres: In Search of Sustainability*. Development Gateway—ICT for Development.

Proenza, F. (2001) Telecenter Sustainability—Myths and Opportunities, in: Dixon and Wattenbach (Eds.) *Bridging the Rural Knowledge Gap: Information Systems for Improved Livelihoods,* www.fao.org/Waicent/FAOINFO/AGRICULT/ags/Agsp/pdf/ProenzaTelecenter.pdf.

Roman, R. and Colle, R. (2002) *Themes and Issues in Telecentre Sustainability*, Development Informatics—Working Paper Series, 10.

Stoll, K. (2003) *Telecentres Sustainability: What Does It Mean?*, Development Gateway—ICT for Development.

The Government of Azerbaijan and UNDP Azerbaijan (2003) *National Information and Communication Technology Strategy 2003–2012*. Baku, Azerbaijan.

UNDP (2004) Azerbaijan *Human Development Report 2003*, Baku, Azerbaijan.

United Nations Economic Commission for Europe (2003) *Towards a Knowledge-Based Economy— Azerbaijan Country Readiness Assessment Report*. Geneva and New York.

About the Author

Michele Cocchiglia is a specialist in ICT for Development policies and projects. He holds a Masters degree in Analysis, Design and Management of Information Systems from the London School of Economics, and a Bachelor degree in Economics from Bocconi University, Italy. His interests include community development and universal access, the socio-economic impact of ICT applications and services, and international politics and economics.

Michele is currently employed within the "E-Government for Development" Technical Unit of the Government of Italy, where he works as a Project Manager for the Balkan Region. Prior to joining the Technical Unit, Michele has served as an ICT consultant for the United Nations Development Programme (UNDP) in Azerbaijan.

Appendix

Regional Information Centres Map of Azerbaijan (January 2004)

Applications of Information Systems in Developing Countries

Introduction

Gert-Jan de Vreede

University of Nebraska at Omaha
College of Information Science & Technology
Department of Information Systems & Quantitative Analysis
phone: +1.402.554-2026
fax: +1.402.554-3400
gdevreede@mail.unomaha.edu

To further the knowledge in the area of Information Systems in developing countries, studies on actual IS applications play a central role. Indeed, accounts of and experiences with particular applications can be educational, inspiring, and even provocative for researchers and practitioners around the world. Moreover, insights on particular classes of applications can shed light on key lessons and guidelines that other researchers and practitioners can learn from and build on. The purpose of this section is to provide a varied yet representative panorama of real information systems applications in developing nations. The chapters touch on issues of information systems development, acceptance, adoption, and use of real applications. Together the chapters in this section illustrate the richness of the research and field work going on in this area.

This section start with "The Roles of Managers in IT Adoption in Thailand—A Case Study of Thai Agricultural Co-operatives" by Oran Chieochan, Theerasak Thanasankit, and Brian Corbitt. The authors note that the development of technically sound applications in itself is not sufficient to ensure use. An important driver for the adoption and subsequent use of Information Technology (IT) is the role and attitude of managers. They examine this phenomenon in the context of Thai agricultural co-operatives that could use IT to improve administrative efficiency. The authors contend that the role and attitude of managers in agricultural co-operatives is strongly influenced by the local cultural, in their case the Thai culture. Cultural value and beliefs can have a considerable influence on IT implementation and adoption processes. Chieochan and colleagues put forward a fourfold analysis framework on IT adoption consisting of individual-level factors, organization-level factors, environmental factors, and characteristics of the innovation itself. These factors were examined through a questionnaire instrument that was filled out by over 250 respondents. The authors found that IT use is more likely to occur if managers are more knowledgeable about IT, are more aware of technology issues, are perceptive towards the social, cultural, and economic conditions, have a propensity to innovative, have a positive attitude towards the use of IT, and are aware of and perceptive towards governmental policies. Although some findings appear to be specifically induced by the Thai culture, many of them appear to be very similar to those found in other developing countries.

The next chapter, "Online Success in a Relationship-Based Economy—Profiles of E-Commerce in China" by Maris Martinsons addresses critical success factors behind E-Commerce initiatives in China. Like their counterparts in other (developed) countries, many Chinese businesses attempt to increase market share and revenue through developing and implementing E-Commerce initiatives. Although often used as a source of inspiration and guidance, Western models of E-Commerce appear to be very unsuccessful in China. The author addresses this issue from a theoretical perspective on relationship-based commerce. This perspective suggests that relationships rather than rules are the prevailing foundation for business in China and helps to explain the difficulties of transferring Western management and organization E-Commerce practices. The author embarked on a longitudinal multiple case study, in which he investigated a number of success stories on online business development in China. His findings offer support for the notion that institutional factors rather than cultural factors are drivers behind relationship-based forms of E-Commerce in China. The common success factor in each case appears to be the ability to overcome institutional deficiencies by building and maintaining strong relationships. In this sense, successful Chinese E-Commerce initiatives appear to resemble electronic networks of co-operating complementary product/service providers, more than electronic markets. Martinsons' insights inform the development of Chinese E-Commerce business models and supporting technical infrastructures. In addition, they set out the special conditions that have to be nurtured through governmental policies and incentives in order to cherish and further stimulate economic development through E-Commerce in China.

The third chapter, "Internet-Based Reverse Auctions by the Brazilian Government—Some Conclusions from a Case Study" by Luiz Antonio Joia and Fuad Zamot addresses challenges and opportunities for e-government in developing countries. Many governments, both in developed and in developing nations, have recognized the potential of IT to bring about fundamental changes in the nature and quality of the services they can deliver to its citizens. E-government initiatives may not only change the way in which governments provide services—indeed they may change the very role of the government in relation to society with respect to transparency, democratic procedures, and citizenship participation. The authors of this chapter explore this phenomenon through a single case study on e-procurement support by the Brazilian Federal Government. They investigate a reverse auction staged by the Federal Government for the purchase of pharmaceutical products by the Ministry of Social Security based on a buy-side centralized procurement management model. In particular, the authors illustrate how such an e-procurement initiative can improve the procurement process in terms of cost reductions, efficiency, efficacy, and accountability. In addition, they present a framework of structural, human, and technical obstacles that may be encountered during the implementation of e-government initiatives, argue which are the underlying causes for these obstacles, and propose how they can be overcome.

The final chapter of this section, "Decision Support Systems for Small Scale Agro-industrial Investment Promotion in Rural Areas" by Carlos Arthur da Silva and Aline Fernandes provides an in-depth account of technological support for decision making in an economic development context. The authors operate in the area of agroindustrial projects that represent investment opportunities. Given the wide-spread belief that agroindustrial projects are a key driver behind rural development, it is important for governments of developing nations to make relevant information on agroindustrial investment opportunities available so that potential investors can make informed decisions. The authors illustrate how the concept of an agroindustrial profile can be generated dynamically through a Decision Support System (DSS), offering timely, accurate, and situation specific information. In co-operation with two Brazilian Ministries, they developed a specific DSS for a substantial number of agroindustrial projects. The DSS has been made available to a wide range of potential users. The authors' findings suggest that the DSS is not only perceived as useful and effective, but that the general concept behind the system should be applied to a wider range of agroindustrial projects. In this sense, the authors' work can serve as an inspiring example for other developing countries of how information technology applications can stimulate rural development initiatives.

9. The Roles of Managers in IT Adoption in Thailand—
A Case Study of Thai Agricultural Co-operatives

Oran Chieochan

Faculty of Science
Naresuan University, Thailand
oranc@nu.ac.th, oran_c2515@hotmail.com

Theerasak Thanasankit

College of Management
Mahidol University, Thailand
jua@tezza.com

Brian J. Corbitt

School of Information Systems
Deakin University, Melbourne, Australia
School of Management, Shinawatra University, Thailand
bcorbitt@deakin.edu.au

Abstract

This paper describes a study of the roles of managers in the adoption of Information Technology (IT) in Thai agricultural co-operatives. These co-operatives are more likely to use IT if they have a manager who has a positive attitude towards IT, is knowledgeable about IT, has a propensity to innovate, and is aware of national economic conditions, government policies, social, cultural and technological issues. Each of the characteristics of managers is explored and the interpretation is enriched by applying a cultural differences framework to gain a better understanding of the influence of each characteristic. The roles of managers are explored in terms of decreasing uncertainty of technology adoption. The role of strong and competent leadership for both the government and the managers is explored in order to gain insights into its correlation with the adoption rate of IT. The paper demonstrates that Thai culture influences the level of IT adoption in Thailand. The paper demonstrates that, on the surface, the managerial characteristics that govern the level of IT adoption in Thailand may be similar to the managerial characteristics demonstrated in the IT adoption literature. The hidden elements (indigenous culture and values) are different, vary from country to country, and influence the adoption factors in a unique way. This uniqueness is dependent on the society where IT is adopted.

1. Introduction

Thai agricultural co-operatives are considered to be large organizations operative in a substantial part of the marketing infrastructure in Thai agriculture. This is significant as Thai agricultural products are one of the major exports of Thailand and contribute significantly (10–15%) to GDP (CPD, 1993, 1999). Agricultural co-operatives are a type of "co-operative society", a term defined by the Macquarie Dictionary as "a business undertaking owned and controlled by its members, and formed to provide them with work or with goods at advantageous prices" (Macquarie Library, 1981). According to the Co-operatives Promotion Department of the Thai Ministry of Agriculture and Co-operatives (CPD, 1993), Thai Agricultural Co-operatives are co-operative only in the sense that they bring together sets of farmers into a bureaucratically controlled entity. They are not managerial co-operatives in the normal sense. Thai agricultural co-operatives perform four main activities—provision of credit for the purchase of agricultural supplies and farm machinery; supply and sale of agricultural necessities and consumer goods; marketing of agricultural produce through local markets, provincial and national federations and abroad; and provision of agricultural extension services. In recent years, IT has been promoted as a tool that Thai Agricultural Co-operatives could use to improve administrative efficiency. The Royal Thai Government has a national plan to support this and has established policy and infrastructure support to enable more widespread use of IT in state organizations, including the agricultural co-operatives.

Thong and Yap (1995) argue that the use of IT by organizations tends to focus on organizational characteristics, without giving due emphasis to the characteristics of individuals. They identify the significant role of an organization's decision-makers in the use of IT. This paper will extend the work of Thong and Yap and argue that managers of Thai agricultural co-operatives influence the use of IT in their co-operatives.

Thanasankit (1999), Jirachiefpattana (1996a and1996b) and Rohitrattana (1998) found that Thai culture has an influence on the use, adoption and development of information systems and technology. This chapter explores the role of Thai culture in the management of IT, applying it to the arguments proposed by Thong and Yap (1995), Yap *et al.* (1994) and Thong (1996, 1999), which have documented IT adoption in Southeast Asia. That research suggests that the variables affecting IT adoption in developed nations may be different from those in developing nations. Similarly, Robey and Rodriquez-Diaz (1989) assert that culture can impede the implementation of information systems because of differences in the way the systems are interpreted and understood. Rohitrattana (1998) also found that indigenous culture and values influence the implementation process of information systems in Thailand.

IT has been used in both developed and developing countries to support operational, tactical and strategic processes within organizations (Abdulgader and Kozar,

1995) and some organizations are faster to introduce IT than others. Many developing countries, notably Thailand, have been investing heavily in IT infrastructure so as to encourage more up take of IT. However, the up take of IT in Thailand is considered to be slow, especially in the agricultural sector (Chiochan and Lindley, 1999). This may be a result of the influence of Thai culture on the manager in Thai agricultural co-operatives.

2. Characteristics of Managers of Thai Agricultural Co-operatives and IT Adoption—A Framework for Analysis

The Thai agricultural co-operative is a bureaucracy which shares certain structural features with business organizations (Laudon and Laudon, 1999). The Thai agricultural co-operative is a social system according to Rogers' (1983, p. 24) definition: "a set of interrelated units that are engaged in joint problem solving to accomplish a common goal". Chiochan (2002, p. 151) suggests that Thai agricultural co-operatives are highly centralized with the main decision maker being the manager. The style of decision-making in Thai agricultural co-operatives would be classified by Rogers (1983) as "authority innovation-decision"[22]. The decision-making processes of Thai agricultural co-operative managers are intuitive, based on guesswork and independent of more formal decision-making models. These are characteristics similar to those proposed by Thong (1996).

Culturally determined concepts, such as belonging, influence interpersonal interaction and relationships, as well as the way in which co-operative business is conducted in Thailand. The adoption of IT is an interactive process, with managers leading the process and receiving support from their employees. Thai decision making does not use a team-based approach, as in Western countries or in Japan (Thanasankit, 1999). From a recent survey in Thailand, it was found that leaders in Thai organizations accept that they have to make decisions in an "authoritarian"[23] way (Holmes and Tangtongtavy, 1995). Authoritarian managers are entitled to make decisions about what they think is correct (Hebert and Benbasat, 1994). It is their job to guide their employees given their role as a father Figure. In Thai agricultural co-operatives, members (farmers) are supposed to participate in the process of decision making but in reality they do not. Farmers are not well educated (Sataporn, 1997) and often feel uncomfortable in the process of decision making, particularly in dealing with IT knowledge. In order to avoid uncertainty in Thailand, there is a tendency to avoid making decisions, since these may bring unwanted tasks and responsibilities.

---

[22] Authority innovation-decisions are choices to adopt or reject an IT that are made by relatively few individuals in a system who possess power, status, or technical expertise (Rogers, 1983).

[23] There is a fine line between an "authoritarian superior" and a "dictatorial manager". A dictatorial manager makes decisions without consulting anyone, but an "authoritarian manger" should nevertheless ask for employees' opinions and show some interest in their views (Hebert and Benbasat, 1994).

Likewise, unsuccessful implementation of a decision may risk job security and invoke the disapproval of employees.

According to Thai government regulations, managers of agricultural co-operatives are responsible for decision-making, managing and administering their co-operatives, including managing budgets, planning and staffing (DEA, 1985). This means that they have a higher social status and are invariably more innovative than their employees. The opinion of managers[24] can influence officers' and members' attitudes towards IT.

Yap (1986) explains that it is difficult to establish the true relationship between IT adoption and organizations. He suggests that organizational factors may determine the use of IT, or the use of IT may influence an organization, or some combination of each. Other theories and models have been used by researchers to explain the inter-relationship of IT and organizations. These include the political conflict model, the organizational ecology model, the managerial innovation model (Robey and Zmud 1992) and the diffusion of innovations model (Rogers, 1983 and 1995). Thong and Yap (1995) and Thong (1999) note that the use of IT is a form of technological innovation, and Lakhanpal (1994, p. 41), in summary, suggests that there are then four categories of relevant factors influencing IT adoption:

(i)   Individual-level factors

(ii)   Organization-level factors

(iii)   Environmental factors

(iv)   Characteristics of the innovation itself

The individual level factors in the Lakhanpal (1994) framework relate to managers and are of particular interest in this paper. Thong and Yap (1995, p. 431), in a study of individual factors within small businesses in Singapore, investigated three influences of chief executive officers (their attitude toward adoption of IT, their IT knowledge, and their innovativeness). Thong and Yap (1995) argue that individual factors can exert a significant influence on the success of information systems within organizations. These conclusions are supported by Fink (1998), Chiochan and Lindley (1999), and Chiochan *et al.* (2000a, 2000b). They also note that a number of other factors, such as national economic conditions, social and cultural contexts, government policy, and technological issues, have not been studied, even though it seems reasonable that they might influence the use of IT in organizations. Based on this literature, we propose a new model which incorporates the managerial issues noted by Thong and Yap (1995) into a cohesive framework with the work of Chiochan and Lindley (1999) and Chiochan *et al.* (2000a, 2000b). This model suggests that the use of IT in Thai agricultural co-operatives relates to the characteristics of managers, influenced by organizational, social, cultural and technological factors (Figure 9.1).

---

[24]   The opinion leadership is explained by Rogers (1983) as the degree to which an individual is able to influence other individual's attitudes or overt behaviour informally in a desired way with relative frequency.

Figure 9.1
Conceptual Research Model of Characteristics of Managers
Affecting IT Adoption in Thai Agricultural Co-operatives

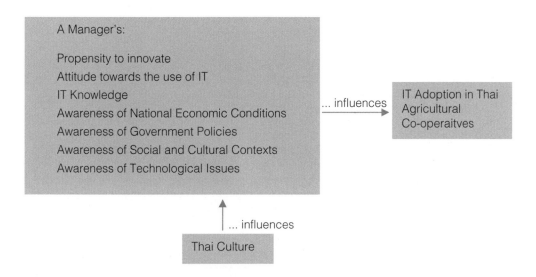

In this research, seven characteristics of managers of Thai agricultural co-operatives were studied—their propensity to innovate, their attitude to the use of IT, their IT knowledge, their awareness of national economic conditions, their awareness of government policies, their awareness of social and cultural contexts, and their awareness of technological issues.

This research sought answers to the following question:

- How does Thai culture influence managers and thus influence the adoption of IT in Thai Agricultural Co-operatives?
- Using existing research, a set of seven hypotheses were derived (Table 9.1), set within the research model described above.

3. Research Methodology

The data in this study was collected from a survey of managers of all Thai Agricultural Co-operatives. The dependent variable in this research is IT adoption, which is measured in two categories. The first is likelihood of use and the second is the extent of use (cf. Yap, 1986, p. 71). Likelihood is measured as a dichotomy—whether the co-operative was computerized or not computerized. This measurement has also been used elsewhere in diffusion research (Fichman, 1992). Alpar and Reeves (1990) have suggested that a business is defined as computerized if it uses at least one major software application such as accounting, sales and purchasing (excluding word processing packages). The independent variables relate to the characteristics of managers and are summarized in Table 9.2.

Table 9.1
Research Hypotheses and Associated Findings from the Literature

H1	Thai agricultural co-operatives with managers who have a propensity to innovate are more likely to use IT.	Thong and Yap (1995) and Thong (1996 and 1999) found that firms are more likely to adopt IT when their CEOs are more innovative.
H2	Thai agricultural co-operatives with managers who have a more positive attitude towards the use of IT are more likely to use IT.	Lakhanpal (1994) found that a middle manager's attitude towards microcomputer usage had a relatively positive influence on the adoption of microcomputers. Also, Thong and Yap (1995) assert that CEOs who possess a positive attitude towards IT are more likely to adopt IT in their business.
H3	Thai agricultural co-operatives with managers who are more knowledgeable about IT are more likely to use IT.	Boynton et al. (1994) found that managerial IT knowledge is a dominant factor in explaining the high level of IT use in organizations. Similarly, Cragg and King (1993) note that the strongest factor inhibiting firms from adopting is a lack of owner knowledge.
H4	Thai agricultural co-operatives are influenced to use IT if managers are aware of national economic conditions.	NSTDA (1997) reported that the economic crisis in Thailand affected not only the government sector, but also the business sector. This means that any kind of business in Thailand, including agricultural co-operatives, might be affected by an economic crisis.
H5	Thai agricultural co-operatives are influenced to use IT if managers are aware of government policies.	Robey and Rodriguez-Diaz (1989) reported that government is one of the factors that inhibits IT in Latin America. For example, unstable government, excessive concern for secrecy and security, priorities are changed frequently, foreign intervention is a constant threat, centralization of political decision making, impact of the scientific approach is minimal at the highest decision levels.
H6	Thai agricultural co-operatives are influenced to use IT if managers are aware of the social and cultural context.	English is the dominant language for development of IT. It is the second language in most Asian countries. Ang and Loh (1998) and Coeur (1997) noted that language is one of the most significant barriers to the use of IT in developing countries.
H7	Thai agricultural co-operatives are influenced to use IT if managers are aware of technological issues.	NSTDA (1997) reported that the communication infrastructure in rural areas of Thailand is still poor and needs improvement.

Table 9.2
Variables in the Study of Factors Affecting the Use of IT
in Thai Agricultural Co-operatives

Independent Variables	Measurement	References
Manager's propensity to innovate	Innovative Not innovative	Kirton (1976)
Manager's attitude to the use of IT	Positive attitude toward use of IT Negative attitude toward use of IT	Moore and Benbasat (1991)
Manager's IT knowledge	More knowledge Less knowledge	Thong and Yap (1995), Thong (1996), Thong (1999)
Manager's awareness of national economic conditions	Perception and awareness of managers to the economy	Johnson and Scholes (1993), Fry and Stone (1995), Hunger and Wheelen (1996), Coulthard et al. (1996)
Manager's awareness of government policies	Perception and awareness of managers to government policy	Johnson and Scholes (1993), Fry and Stone (1995), Hunger and Wheelen (1996), Coulthard et al. (1996)
Manager's awareness of social and cultural contexts	Perception and awareness of managers to the social and cultural context	Johnson and Scholes (1993), Fry and Stone (1995), Hunger and Wheelen (1996), Coulthard et al. (1996)
Manager's awareness of technological issues	Perception and awareness of managers to infrastructure and technology	Rogers (1983), Johnson and Scholes (1993), Fry and Stone (1995), Hunger and Wheelen (1996), Coulthard et al. (1996)

Data was collected by a questionnaire which was mailed to managers of Thai agricultural co-operatives. Following Sarantakos (1998), the questionnaire used a five-point Likert scale with a range of responses from "strongly disagree" to "strongly agree". The questionnaires were pre-tested using a semi-structured interview—enabling discussion of complex or ambiguous questions. Sarantakos (1998) and Lawrence and Keen (1996) recommended this as a way of improving question wording. During the pretest, three types of people were interviewed—the head of the IT Section, Ministry of Thai Agricultural Co-operatives; two Thai researchers; and five Thai agricultural co-operative managers. This data was used later to confirm the analysis resulting from the survey. This data was collected in an interpretivist framework and represents the views and rich meanings of some managers of Thai Agricultural Co-operatives. To some extent the source interpretive data is limited but the findings were so consistent across all those interviewed that the reflections in the transcribed data are considered to be of important value in understanding what is happening in Thai Agricultural Co-operatives.

The modified questionnaires were then evaluated in a pilot test by sending them to 20 randomly selected managers of Thai agricultural co-operatives. Fourteen questionnaires were returned, including 4 that were incomplete.

The names and addresses of Thai agricultural co-operatives were obtained from the Co-operatives Promotion Department, Ministry of Thai Agriculture and Co-operatives. There are 3,398 Thai agricultural co-operatives with 4,659,373 members. The criterion for determining the size of Thai agricultural co-operatives is based on the number of members (farmers). Srisempook, (1993) classified Thai agricultural co-operatives into three sizes. The largest size has more than 2,000 members, medium size has between 1,000 and 2,000 members and the smallest size has fewer than 1,000 members.

A multistage sampling technique was used, namely systematic random sampling and stratified random sampling. Sproull (1995) noted that a 10% sample is adequate. A sample of 454 co-operatives was drawn and 256 useable responses were obtained giving a 56% response rate. The data was analyzed in two stages. First, two bivariate data analyses (t-test and Pearson correlation) were used for testing individual hypotheses relating to the likelihood of IT usage. Pearson correlation was used to test individual hypotheses relating to extent of IT usage. This first stage of data analysis used a t-test analysis to compare the differences of means between two groups—Thai agricultural co-operatives using and not using IT.

Secondly, a multivariate analysis (discriminant analysis) was used to study the effects of all independent variables simultaneously on the likelihood of IT usages. Multiple regression was used to examine of the effects of the independent variables on the extent of IT usages. This second stage of data analysis used discriminant analysis to identify the combination of independent variables which best account for the statistically significant differences between Thai agricultural co-operatives that use or do not use IT in their operations. Multiple regression analysis is used for prediction and explanation and for building a structural model to determine the extent to which each predictor variable affects the outcome variables.

4. Data Analysis and Discussion

In general, the results of the survey analysis support the conclusions of Thong and Yap (1995) and Thong (1996, 1999). It can be seen in Tables 9.3 and 9.4 that the manager's knowledge about IT tested at the highest level of significance at 0.000 (T value=11.685 and 10.127, Correlation coefficient=0.594 and 0.537 and F value=116.313 and 89.613). Each of the other variables shows weaker T-values but the data is a significant explanator of the likelihood of using IT in the Thai agricultural co-operatives.

Table 9.3
The Result of T-test to Investigate the Likelihood of Using IT
in Thai Agricultural Co-operatives

Variables (Manager Characteristics)	Computerized (n=153)		Non-computerized (n=103)		T-test	
	Mean	SD	Mean	SD	T-value	2-tailed. SIG
H1a: Manager's propensity to innovate	3.9208	0.3603	3.7782	0.3958	2.983	0.003**
H2a: Manager's attitude to the use of IT	3.9564	0.2857	3.8314	0.4106	2.872	0.004**
H3a: Manager's IT knowledge	3.4211	1.2044	1.6832	1.0856	11.685	0.000**
H4a: Manager's awareness of national economic conditions	3.5490	0.6865	3.6238	0.6510	3.404	0.001**
H5a: Manager's awareness of government policies	3.3170	0.7418	3.1197	0.7462	2.081	0.038*
H6a: Manager's awareness of Social and cultural contexts	3.9069	0.6570	3.6044	0.7561	3.398	0.001**
H7a: Manager's awareness of technological issues	4.1786	0.6283	3.9288	0.7693	2.848	0.005**

*Significant at p < 0.05, **significant at p<0.01*

Table 9.4
Pearson Correlation Investigating the Likelihood of Using IT in
Thai Agricultural Co-operatives

Variables (Manager Characteristics)	Likelihood of the Use of IT	
	Correlation Coefficient	Significance
H1a: Manager's propensity to innovate	0.184	0.003**
H2a: Manager's attitude to the use of IT	0.177	0.004**
H3a: Manager's IT knowledge	0.594	0.000**
H4a: Manager's awareness of national economic conditions	0.209**	0.001**
H5a: Manager's awareness of government policies	0.129*	0.038*
H6a: Manager's awareness of Social and cultural contexts	0.209**	0.001**
H7a: Manager's awareness of technological issues	0.176	0.005**

*Significant at p < 0.05, **significant at p<0.01*

In the measured relationships between the likelihood of using IT and the independent variables (Table 9.4), the relationships are weak except for the manager's attitude towards IT. The strength of this variable as an explanator of the likelihood of Thai agricultural co-operatives using IT is illustrated further in Table 9.5 with the strength of the F value and the strong measure in the canonical correlation coefficient.

Table 9.5
Discriminant Analysis Investigating the Likelihood of Using IT
in Thai Agricultural Co-operatives

Variables (Manager Characteristics)	Univariate F-test Probability	Significance	Standard-ized Canonical Coefficients	Structure Matrix: Pooled within-groups Correlations
H1a: Manager's propensity to innovate	8.570	0.004**	0.085	0.171
H2a: Manager's attitude to the use of IT	9.384	0.002**	0.259	0.179
H3a: Manager's IT knowledge	116.313	0.000**	0.622	0.630
H4a: Manager's awareness of national economic conditions	11.591	0.001**	0.244	0.199
H5a: Manager's awareness of government policies	4.372	0.038*	0.160	0.122
H6a: Manager's awareness of Social and cultural contexts	15.558	0.000**	0.263	0.231
H7a: Manager's awareness of technological issues	10.156	0.002**	0.019	0.186

*Significant at $p < 0.05$, **significant at $p < 0.01$

The extent of IT use in the Thai agricultural co-operatives appears to be more strongly associated with the manager's awareness of government policy than other indicators (Table 9.6). The coefficient data suggests that the relationships reported in the hypotheses are supported.

Table 9.6
Pearson Correlation Investigating the Extent of Use of IT
in Thai Agricultural Co-operatives

Variables Manager Characteristics)	Pearson Correlation	
	Correlation Coefficient	Significance
H1b: Manager's propensity to innovate	0.013	0.869
H2b: Manager's attitude towards the use of IT	-0.011	0.897
H3b: Manager's IT knowledge	-0.068	0.408
H4b: Manager's awareness of national economic conditions	0.107	0.189
H5b: Manager's awareness of government policy	-0.069	0.175
H6b: Manager's awareness of social and cultural contexts	-0.079	0.334
H7b: Manager's awareness of technological issues	-0.053	0.518

*Significant at $P < 0.05$, ** Significant at $P < 0.01$*

This is supported in the multiple regression analysis both in the coefficients reported and with T values (Table 9.7).

Table 9.7
Multiple Regressions Coefficients Indicative of the Extent of Use of IT
in Thai Agricultural Co-operatives

Variables (Manager Characteristics)	Standardized Coefficient Beta	T-value	Significance
H1b: Manager's propensity to innovate			
H2b: Manager's attitude towards the use of IT	0.029	0.316	0.753
H3b: Manager's IT knowledge	0.031	0.364	0.716
H4b: Manager's awareness of national economic conditions	0.105	1.169	0.254
H5b: Manager's awareness of government policy	-0.031	-0.363	0.717
H6b: Manager's awareness of social and cultural contexts	-0.040	-0.437	0.663
H7b: Manager's awareness of technological issues			

$R=0.462$, $R2=0.214$, Adjust $R2=0.110$, $F =2.064$, $P =0.012$

*Significant at $P < 0.05$, ** Significant at $P < 0.01$*

Following the methodology of Morgan and Griego (1998), the analysis suggests that all hypotheses (H1a–H7a) are supported, but only to a weak level. The results of the extent of the use of IT analysis (H1b–H7b) indicate that no hypothesis is supported in terms of significance. It can therefore be argued that the characteristics of managers of Thai agricultural co-operatives appear not to influence the extent of IT use in Thai agricultural co-operatives.

If this is the case then it is important to explain what is happening in the Thai agricultural co-operatives to understand why IT adoption is not as influenced by managers to the same extent as it was in previous studies. To understand this we argue that the inferred nature of Thai culture must be used to explain the behaviour of managers in the adoption of IT in Thai agricultural co-operatives.

4.1 Knowledge about IT

Managers' knowledge about IT appears to be the most important factor in the use of IT in Thai agricultural co-operatives. This finding is consistent with Attewell's (1992) theory of lowering knowledge barriers, which implies that business IT is not just a matter of purchasing objects (computers and software) but requires considerable organizational learning, skill development and overcoming knowledge barriers.

The concept of overcoming uncertainty in Thailand has a significant impact on the level of confidence of a manager's subordinates towards the organizational adoption of IT. Thai subordinates tend to pass decision-making tasks to their superiors. During the process of adopting, many decisions have to be made such as: selecting the type of hardware and software; selecting tasks for automation; and reengineering organizational processes. As Holmes and Tangtongtavy (1996, p.63) indicate, subordinates trust their manager to:

- … decide things. Since he has qualified as a boss, it is assumed that he possesses certain knowledge, wisdom, or experience which go beyond the capacity of his colleagues.

The level of IT knowledge possessed by managers has an impact on the level of uncertainty during the adoption of IT. When managers in Thai agricultural co-operatives decide to exploit IT in their branches, their subordinates expect their managers to understand how to use and develop IT appropriately (cf. also Holmes and Tangtongtavy, 1996). Such an understanding means that managers can explain IT-related issues to subordinates, at the same time preserving "face". Consequently, the level of managerial IT knowledge in Thai agricultural co-operatives influences the level of support they receive from their subordinates and is directly associated with the successful adoption of IT. Even if employees possess a higher level of IT knowledge than their managers, they tend not to demonstrate their knowledge during meetings so as to preserve the manager's face. Indeed, Komin (1990: p. 135) suggests that:

- …"face" is identical with "ego" and is very sensitive. Since the Thais give tremendous emphasis on "face" and "ego", preserving one another's "ego" is the basic rule of all Thai interactions both on the continuum of familiarity-unfamiliarity, and the continuum of superior-inferior, with difference only in degree.

Whenever there is a problem to be solved, Thais tone down the negative messages used, avoiding confrontation in public such as during meetings. They try to avoid making a person lose face at all costs, since causing another person to lose face is a form of insult. Komin (1990: p. 135) suggests that Thais "are very 'ego' oriented, to the extent that it is very difficult for the Thai to dissociate one's ideas and opinions from the 'ego' self". This is why a person whose ideas are criticized will take the criticism personally and not as criticism of the ideas themselves.

Employees tend not to challenge the managers' ideas as they feel *Kreng Jai* and want to preserve their managers' face. *Kreng Jai* (considerate) is defined by Klausner (1981: p. 199), as:

- ... to be considerate, to feel reluctant to impose upon another person, to take another person's feelings (and "ego") into account, or to take every measure not to cause discomfort or inconvenience for another person.

Kreng Jai behaviour can be observed by all superiors, equals, and inferiors and includes intimate relationships between husband and wife, as well as between close friends. The only difference is the degree of *Kreng Jai*. Komin (1990: p. 136) suggests that:

- A Thai knows how far he/she can go in displaying the degree of *Kreng Jai* in accordance to different degree of status discrepancy, degree of familiarity, and different situations.

Showing *Kreng Jai* towards one who is higher in rank and seniority also means showing consideration merged with respect, especially showing support for the manager's leadership. Challenging a manager's ideas also implies a lack of respect (cf. Thanasankit, 1999 and Rohitratana, 1998). By avoiding conflict and challenging ideas, managers and their employees preserve their good relationships, another important element in Thai culture. Without confidence in their knowledge, Thai managers will rarely adopt IT in their organizations. IT is a risk element for managers. To avoid conflict with users, managers in the Thai context would rather avoid adoption than create problems. To be sure of their status, IT knowledge is fundamental for these managers. Without that knowledge, the managers practice avoidance.

4.2 Manager's Awareness of the Technology and Perceptiveness towards Social, Cultural and Economic Conditions

In general, Roger's (1995) theory helps explain adoption of IT in terms of its characteristics as an innovation. One explanation of the results in this study is that managers of Thai agricultural co-operatives are more concerned with inherent IT characteristics (i.e., relative advantage, compatibility, complexity, trialability, and observability), than they are with external factors such as research and development, government and industry support and existing infrastructure.

Social and cultural factors can also affect the use of IT. In this study, social and cultural factors such as the English language barrier, and the attitude towards the quality of working life (Johnson and Scholes, 1993; Fry and Stone, 1995; Hunger and Wheelen, 1996; Coulthard *et al.*, 1996) were identified by managers. IT can improve the quality of life for a manager and his/her employees and help people finish their jobs faster and more easily than with old methods. However, the mangers need to be aware of how and what IT to use in their organizations. Odedra (1993) suggests that some cultures might perceive inappropriate design and use of IT as a disruption to their communication and interaction processes.

Managers in the Thai agricultural co-operative needed to understand the employees' attitude towards IT. Their employees may believe, for instance, that the IT will replace their jobs. This is likely to result in the development of resistance to IT adoption. The concept of belonging can be used to illustrate the situation where the security of their colleagues' jobs was perceived to be threatened by IT.

In 1997, economic conditions in Thailand were difficult (Terdudomtham, 1998) and the number of co-operatives using computers decreased from 1997 to 1999. The managers in Thai agricultural co-operatives needed to keep up with the economic conditions as incorrect predictions and investments might increase the cost of IT. This would then decrease the level of employee confidence in their knowledge about the economy, which might then cause the managers to lose face. Managers in this study were keenly aware of the technology required and of the impact of economic conditions to the extent that they modified and adapted their behaviour with IT adoption to those conditions. The data from the survey highlights that this issue is important but its strength relative to other variables as an explanatory of IT adoption is weaker, as perceived by the managers themselves.

4.3 Managers Have a Propensity to Innovate

The managers in Thai agricultural co-operatives are required to be creative in the use and adoption of IT. Innovative mangers are expected to be able to select and adapt IT to be used in their organizations. IT is typically developed in the West and relies on a Western ideology (Odedra, 1993). Sometimes that IT needs to be changed or modified, generally with local expertise, to suit Thailand's economic, social and political climate.

A manager's ability to solve IT problems creatively and innovatively results in increased trust and confidence from employees, since these managers can reduce uncertainty levels for the adoption of IT. Employee trust and a good relationship between mangers and their subordinates will benefit the IT adoption process. Relationship-orientation behaviour happens more commonly than work-oriented behaviour, both within Thai society and Thai organizations (Thanasankit, 1999). The lack of close relationships between management and subordinates has been shown to influence Executive Information Systems development in Thailand (Jirachiefpattana, 1996a and 1996b) and it is not unreasonable to suggest that it also affects IT adoption in Thai agricultural co-operatives.

The manager's innovativeness was perceived to be an individual characteristic, which may not be directly related to Thai culture from the data. However, the result of innovation would relate to the manager's leadership and knowledge, which were important considerations in terms of gaining employee respect and trust, and so encouraging employees to adopt IT. The use of innovative solutions that help employees adopt IT enables managers to build *Bun Khun,* which is:

- ... indebted goodness, a psychological bond between someone who, out of sheer kindness and sincerity, renders another person the needed help and favour, and the latter's remembering of the goodness done and his ever-readiness to reciprocate the kindness (Holmes and Tangtongtavy, 1995: p. 30).

This *Bun Khun* is used for the base relationship between two people to respect and or "do favours for" each other. *Bun Khun:*

- ... must be returned, often on a continuous basis and in a variety of ways, because *Bun Khun* should not and cannot be measured quantitatively in material terms. It is an ongoing, binding of good reciprocal feelings and lasting relationship... the Thais are brought up to value this process of gratefulness—the process of reciprocity of goodness done, and the ever-readiness to reciprocate. Time and distance are not the factors to diminish the *Bun Khun* (Komin 1990: p. 39).

For subordinates to create conflict or influence their superiors in meetings or public places would be considered as not respecting them and not returning *Bun Khun.* However, it is important to understand that superiors build/create *Bun Khun (Saang Bun Khun)* toward their subordinates in the first instance. Only then can they expect to receive reciprocal gratitude from their subordinates. When subordinates make mistakes in their work or create a negative impression towards clients, their superiors will try to cover up for them. Therefore, not returning *Bun Khun* may sour their relationship and may effect their mutual co-operation. *Saang Bun Khun* can also be exploited and used in establishing power in Thai society (Rohitratana, 1998). This power and connections can build an empire for that person. Subordinates, under their superior's empire, might then gain a certain degree of protection and benefits in return (Komin, 1990; Thanasankit, 1999). This process may create job security for subordinates and good relationships with their superiors. They will work in a group and share their success.

The concept of *Bun Khun* plays an important role throughout all levels in the Thai social hierarchy. *Bun Khun* creates a behavioural pattern, which enables different levels of people at different social levels to interact in a friendly manner (Holmes and Tangtongtavy, 1995). *Bun Khun* would thus exert a significant impact on the adoption of IT in the Thai agricultural co-operatives where co-operation from subordinates is essential. By knowing their manager's ability to solve problems creatively, the employees also show *Kreng Jai* where they can increase their respect toward their managers.

These cultural practices of *Kreng Jai* and *Bun Khun* influence the propensity of managers in the Thai agricultural co-operative to be truly careful about adopting IT. There is a demand that they preserve their relationships with their staff and with the farmer members involved. If IT is a challenge to those relationships, then adoption is negatively affected. If, however, there is evidence that IT adoption enhances efficiency and profitability of the co-operative, and this was mostly the case, then IT adoption was embraced. These cultural practices fundamentally drive the attitudes of the managers of the agricultural co-operatives and both enhance and create barriers in professional practice. IT is a new phenomenon for the farmer members and the adoption of IT within the co-operatives is more about alignment with these cultural practices rather than just an economically rationalized defence of decisions.

4.4 Managers Have a Positive Attitude towards the Use of IT

Managers of Thai agricultural co-operatives who are more innovative and have a positive attitude towards the use of IT tend to use IT in their organizations. This finding is consistent with Roger's model of an individual's innovation-adoption process (Rogers, 1983 and 1995). Managers in Thai organizations are perceived as "father Figures" in Thai organizations. When the managers show positive attitudes towards the use of IT, they simultaneously show their employees the benefits and usefulness of IT. This then results in increasing support and co-operation from their employees in the adoption of IT. The managers' attitudes are also reflected in their employee's attitude toward IT. This positive attitude encourages managers to adopt IT to improve their organizations so as to work more effectively and efficiently. By encouraging their employees to work as a team and get the whole organization involved in the implementation of an IT project, the mangers also create a sense of working together as a group and create the value of belongingness for their employees. By working in a group with a positive attitude toward IT, the mangers tend to minimize the level of uncertainty, which would increase the employee's confidence and trust in the use of IT in Thai agricultural co-operatives.

4.5 Managers' Awareness of and Perceptiveness towards Policies from the Royal Thai Government and the Thai Ministry of Agriculture and Co-operatives

Policies from the Royal Thai Government and the Thai Ministry of Agriculture and Co-operatives, such as the development and promotion of free software, have been promulgated in an attempt to influence the use of IT in Thai agricultural co-operatives. In 1995, when the Thai government announced the Year of IT (Koanatakool, 1998), the number of Thai agricultural co-operatives starting to use computers was highest.

Thai organizations, including Thai agricultural co-operatives, are very structured. The policies from the Ministry of Agriculture and Co-operatives are very important to provide direction in the use of IT. Following that direction from the Ministry would reduce the level of the uncertainty for the managers in the use of IT. The assistance from the Ministry, such as free software, also provides the managers with support in selecting the type of software and hardware for their co-operatives. The concept of power played an important role.

The mangers provide their employees with directions for minimizing uncertainty and gaining co-operation and support from their employees. However, the managers also needed direction and support, which comes from the Ministry. The policies from the Ministry are considered as the "golden rule" for the mangers as it is socially accepted that they are required to follow these policies set by the minister and a Thai Civil Service committee, comprised of high profile business persons and academics. A failure to follow these policies would imply a lack of respect for the Minister. Such behaviour is not encouraged or accepted in Thai society. If the managers do not follow the given policies, their employees lack confidence in them and this would both increase the level of uncertainty and decrease the level of trust from their employees. Hofstede (1991) suggests that in countries characterized by a high power distance, individuals prefer to: treat the elders with respect; have centralized control for decision-making; accept that being told what to do by superiors is appropriate; acknowledge the existence of a father Figure within the country or in organizations.

5. Conclusion

This project set out to test theories associated with the adoption of IT in Thai agricultural co-operatives. It also attempted to identify characteristics of managers that could improve industry practice in agricultural organizations in nations like Thailand where there is a need to learn and change. This paper provides preliminary empirical evidence of managerial characteristics affecting IT adoption in Thai agricultural co-operatives.

Thai Agricultural Co-operatives are more likely to use IT when their managers:

- are more knowledgeable about IT
- are aware of technology issues
- are perceptive towards social, cultural and economic conditions
- have a propensity to innovate
- have a positive attitude towards the use of IT
- are aware of and perceptive towards policies from the Royal Thai Government and the Thai Ministry of Agriculture and Co-operatives.

The surface factors are very similar to those found in research conducted in other developing countries (Thanasankit, 1999; Thanasankit and Corbitt, 1999a, 1999b, 2000). However, a variety of explanations can be employed to describe each factor found in this study. One of those explanations involves Thai culture. Thai culture provides this study with a unique perspective in detecting and elucidating the hidden elements that need to be achieved if successful implementation and adoption of IT is to be achieved. The hidden elements are unique to individual indigenous cultures. Therefore, there is a need to study each culture thoroughly (especially when a country has more than one dominant culture) if we are to gain a better understanding of the role of culture in IT management.

There are also a number of limitations associated with this research. Firstly, there may be other theories and frameworks that can be used to describe the research findings rather than employing a Thai culture framework. Other theories that can be used to describe the findings can be the diffusion of innovations theory (Rogers, 1983 and 1995), organizational politics, gender differences, attitude toward IT, and notions of success and acceptance. The study was conducted in a large public organization, where the factors may be unique and vary from private organizations as there are organizational culture differences between those two sectors (Thanasankit, 1999). However, the findings can be used to guide future research with a larger sample size and more diverse industrial sectors.

In a policy context the findings of this research are important because they stress the fine grain of cultural impact. They show that policy makers have to understand the micro level of policy implementations and its impact as much as the macro expectations of national development stressed in broad policy. Farmers in Thai Agricultural Co-operatives related to the managers of those co-operatives in ways that are based on Thai cultural codes such as *Kreng Jai* and *Bun Khun* which establish the nature of the relationships. Broad scale policy reflects whole of nation outcomes such as economic growth and social development which subsume the fine grain of grounded relationships. We would suggest that developing countries are in situations where ideologies and economic and management practices of other cultures are being imposed without a serious understanding of the fine grained practices at the micro level. This chapter has gone some way to addressing that understanding, but realistically, this chapter is only the beginning of that understanding.

References

Abdulgader, A. H. and Kozar, K. A. (1995) The Impact of Computer Alienation on Information Technology Investment Decisions: An Exploratory Cross-National Analysis, *MIS Quarterly,* **19**, 4, 535–559.

Ang, P. H. and Loh, M. C. (1998) Internet Development in Asia, Internet Society (ISOC), info.isoc.org:80/isoc/whatis/conferences/inet/96/proceedings/h1/h1_1.htm,

Alpar, P. and Reeves, S. (1990) Predictors of MS/OR Application in Small Businesses, *Interfaces*, **20**, 2, 2–11.

Attewell, P. (1992) Technology Diffusion and Organization Learning: The Case of Business Computing, *Organization Science*, **3**, 1, 1–19.

Boynton, A. C., Zmud, R. W. and Jacob, G. C. (1994) The Influence of IT Management Practice on IT Use in Large Organizations, *MIS Quarterly*, **18**, 3, 299–318.

Chieochan, O. and Lindley, D. (1999) Factors Affecting the Use of Information Technology in Thai Agricultural Co-operatives: A Work in Progress, *Proceedings of the Conference on Information Technology in Asia,* Kuching, 136–148.

Chieochan, O., Lindley, D. and Dunn, T. (2000a) Factors Affecting the Use of Information Technology in Thai Agricultural Co-operatives: A Work in Progress, *The Electronic Journal on Information Systems in Developing Countries*, **2**, 1, 1–17.

Chieochan, O., Lindley, D. Dunn, T. (2000b) Factors Affecting the Use of Information Technology in Thai Agricultural Co-operatives: A Preliminary Data Analysis, *Proceedings of the 4th Pacific Asia Conference on Information Systems,* Hong Kong, 389–407.

Chieochan, O. (2002) Factors Affecting the Use of Information Technology in Thai Agricultural Co-operatives. *Ph.D. Dissertation*, School of Information Studies, Charles Sturt University.

Coeur, D. R. O (1997) The African Challenge: Internet, Networking and Connectivity Activities in a Developing Environment, *Third World Quarterly*, **18**, 5, 883–898.

Coulthard, M., Howell, A. and Clarke, G. (1996) *Business Planning: The Key to Success*. South Melbourne: Macmillan Education.

CPD (1993) Co-operatives Promotion Department, *Co-operatives in Thailand*. Bangkok: Thai Ministry of Agriculture and Co-operatives.

CPD (1999) Co-operatives Promotion Department, *Annual Report of Co-operatives in Thailand*. Bangkok: Thai Ministry of Agriculture and Co-operatives.

Cragg, P. B and King, M. (1993) Small-firm Computing: Motivators and Inhibitors, *MIS Quarterly*, **17**, 1, 47–60.

DEA (1985) Department of Agricultural Extension, *Thai Agricultural Co-operative Learning*. Bangkok: Ministry of Agriculture and Co-operatives.

Fichman, R. G. (1999) Information Technology Diffusion: A Review of Empirical Research, *Proceedings of the 13th International Conference on Information Systems*, Dallas, 195–206.

Fink, D. (1998) Guidelines for Successful Adoption of Information Technology in Small and Medium Enterprises, *International Journal of Information Management*, **18**, 14, 243–253.

Fry F. L. and Stone, C. R. (1995) *Strategic Planning for the New Small Business.* Dover: Upstart Publishing Co.

Hebert, M. and Benbasat, I. (1994) Adopting Information Technology in Hospitals: The Relationship between Attitudes/Expectations and Behavior, *Hospital and Health Services Administration*, **39**, 3, 369–384.

Hofstede, G. (1991) *Culture and Organizations.* UK: McGraw-Hill.

Holmes, H. and Tangtongtavy, S. (1996) *Working with the Thais.* Bangkok: White Lotus Co.

Hunger D. and Wheelen, T. L (1996) *Strategic Management.* Reading: Addison-Wesley.

Jirachiefpattana, W. (1996a) An Examination of Some Methodological Issues in Executive Information Systems Development in Australia and Thailand. *Doctoral Dissertation*, Department of Information Systems, Monash University.

Jirachiefpattana, W. (1996b) The Impact of Thai Culture on Executive Information Systems Development, *Proceedings of the 6th International Conference on Thai Studies, Theme 1: Globalization: Impact on and Coping Strategies in Thai Society*, Chiang Mai, 97–110.

Johnson, G. and Scholes, K. (1993) *Exploring Corporate Strategy* (3rd ed.), New York: Prentice Hall.

Kirton, M. (1976) Adaptors and Innovators: A Description and Measure, *Journal of Applied Psychology*, **61**, 5, 622–629.

Klausner, W. J. (1981) Reflections on Thai Culture: Collected Writings. Bangkok: Suksit Siam.

Koanatakool, T. (1998) National IT Project in Thailand: IT Project into the Future. National Electronics and Computers Technology Center (NECTEC), www.nectec.or.th/it-projects/index.html.

Komin, S. (1990) *Psychology of The Thai People: Values and Behavioral Patterns*. Bangkok: NIDA (National Institute of Development Administration).

Lakhanpal, B. (1994) Assessing the Factors Related to Microcomputer Usage by Middle Managers, *International Journal of Information Management*, **14**, 1, 39–50.

Laudon, K. C. and Laudon, J. P. (1999) *Essentials of Management Information Systems, Transforming Business and Management*. Upper Saddle River: Prentice-Hall.

Lawrence, K. L. and Keen, C. D. (1996) A Survey of Factors Inhibiting the Adoption of Electronic Commerce by Small and Medium Enterprises in Tasmania, *Working Paper WP 97–01*, Department of Information Systems, University of Tasmania.

Macquarie Library (1981) *The Macquarie Dictionary*. Sydney: Macquarie Library.

Moore, G. C and Benbasat, I. (1991) Development of an Instrument to Measure the Perceptions of Adopting and Information Technology Innovation, *Information Systems Research*, **2**, 3, 192–222.

Morgan, G. A and Griego, O. V. (1998) *Easy Use and Interpretation of SPSS for Windows: Answering Research Question With Statistics*. Mahwah: Lawrence Erlbaum Assoc.

NSTDA (1997) National Science and Technology Development Agency, What Thailand Needs to Survive, www.nstda.or.th/newsroom/pr/pr241097a.html.

Odedra, M. (1993) Enforcement of Foreign Technology in Africa: Its Effect on Society, Culture, and Utilization of Information Technology, in: Beardon, C. and Whitehouse, D. (eds.) *Computers and Society*. Oxford: Intellect.

Robey, D. and Rodriguez-Diaz, A. (1989) The Organizational and Cultural Context of Systems Implementation: Case Experience from Latin America, *Information and Management*, **17**, 4, 229–239.

Robey, D. and Zmud, R. (1992) Research on the Organization of End-User Computing: Theoretical Perspectives from Organization Science, *Information Technology and People*, **6**, 1, 11–27.

Rogers, E.M. (1983) *Diffusion of Innovations* (3rd ed.), New York: Free Press.

Rogers, E.M. (1995) *Diffusion of Innovations* (4th ed.), New York: Free Press.

Rohitratana, K. (1998) The Role of Thai Values in Managing Information Systems: A Case Study of Implementing an MRP System, *Proceedings of the 5th Working Conference of IFIP WG 9.4*, Bangkok, 188–201.

Sarantakos, S. (1998) *Social Research*. Charles Sturt University.

Sataporn, P. (1997) *Co-operative Management*. Bangkok: Odeon Store Pub.

Sproull, N.L. (1995) *Handbook of Research Methods: A Guide for Practitioners and Students in the Social Sciences*. London: The Scarecrow Press.

Srisempook, K. (1993) Factors Affecting the Resignation of Employees in Thai Agricultural Co-operatives. *Masters Thesis*, Thammasat University.

Terdudomtham, T. (1998) The Bubble Finally Burst: Economic Review Year End 1997, Bangkok Post, www.bangkokpost.net/ecoreview97/review9701.html.

Thanasankit, T. (1999) Exploring Social Aspects of Requirements Engineering—An Ethnographic Study of Thai Systems Analysts, *Doctoral Dissertation*, Department of Information Systems, The University of Melbourne.

Thanasankit, T. and Corbitt, B. (1999a) Towards Understanding Managing Requirements Engineering: A Case Study of a Thai Software House, *Proceedings of the 10th Australasian Conference on Information Systems*, Victoria University of Wellington, 993–1013.

Thanasankit, T. and Corbitt, B. (1999b) Towards an Understanding of the Impact of Thai Culture on Requirements Elicitation, *Proceedings of the Conference on Information Technology in Asia*, Kuching, 420–440.

Thanasankit, T. and Corbitt, B. (2000) Thai Culture and Communication of Decision Making Processes in Requirements Engineering, *Proceedings of the 2nd International Conference on Cultural Attitudes Towards Technology and Communication*, Murdoch University, 217–242.

Thong, J. Y. L. (1996) Information Systems Adoption and Implementation in Small Businesses in Singapore, *Doctoral Dissertation,* National University of Singapore.

Thong, J. Y. L (1999) An Integrated Model of Information Systems Adoption in Small Businesses, *Journal of Management Information Systems*, **15**, 4, 187-–14.

Thong, J. Y. L and Yap, C. S. (1995) CEO Characteristics, Organizational Characteristics and Information Technology Adoption in Small Business, *Omega*, **23**, 4, 429–442.

Yap, C. S. (1986) Information Technology in Organizations in the Service Sector, *Doctoral Dissertation,* University of Cambridge.

Yap, C. S., Thong, J. Y. L and Raman, K. S. (1994) Effect of Government Incentive on Computerization in Small Business, *European Journal of Information Systems*, **3**, 3, 191–206.

About the Authors

Dr. Oran Chieochan has a Bachelor of Science (first class honours) degree from Rajamangala Institute of Technology, Thailand, a Master of Computing degree from Griffith University, Australia, and a PhD from the School of Information Studies, Charles Sturt University, Australia. He is presently an Assistant Dean for the Faculty of Science at Naresuan University, Thailand and is a Course Co-ordinator for Master of Information Technology and Bachelor of Information Systems. He is also a freelance researcher for Global Concerns Company Limited and a columnist, for Thai news papers in Thailand. His research and publications focus on IT and electronic commerce in agricultural businesses and rural development in developing countries, especially in Thailand.

Dr. Theerasak Thanasankit is currently the Director of Asia Pacific Advanced Technology Co. Ltd. and holding lectureship at the College of Management, Mahidol University, Thailand. Dr. Thanasankit previously was Lecturer then Senior Lecturer in Information Systems at the University of Melbourne, Victoria University of Wellington and then Monash University. His research and publications focus on requirements engineering and E-Commerce and on the way national cultures differentiate requirements gathering processes. This research is enabling comparative studies in E-Commerce to be compiled to provide assistance to business decision-makers in the adoption of new technologies.

Professor Brian Corbitt is currently Professor of Management Science at Shinawatra University in Thailand and Professor of Information Systems at Deakin University. He has previously been Professor of eCommerce at Victoria University of Wellington in New Zealand. And prior to that lectured at the University of Melbourne, where he was Head of International House, and before that Monash University. He specializes in Electronic Commerce policy development, analysis and implementation and in Business Modelling and Electronic Commerce trade relationships, and knowledge management in tertiary institutions. He has published 4 books on eBusiness and eCommerce, another 4 books, some 74 refereed scholarly papers, numerous government reports to the Governments of Thailand and New Zealand, and some 20 invited papers as a keynote speaker in Malaysia, Singapore, Thailand, New Zealand, Japan, Hong Kong, and Australia. He is currently responsible for the development of E-Learning and online initiatives at Deakin University as Pro Vice Chancellor.

10. Online Success in a Relationship-Based Economy—
Profiles of E-Commerce in China

Maris G. Martinsons
City University of Hong Kong
Kowloon, Hong Kong, China
mgmaris@cityu.edu.hk

Abstract

The author has developed a theory of relationship-based commerce to help explain the difficulties of transferring Western theories and prescriptions for management and organization to countries such as the People's Republic of China (China). This theory suggests that social relationships are the prevailing foundation for business in China. Such a relationship-based foundation differs fundamentally from the rule-based foundation for business that prevails in the United States and other modern market-based economies.

The focus of this chapter is online business development in China. Emerging success stories in the B2B, B2C and C2C marketspaces are profiled. These include an online retailer, one of the world's leading supply chain management firms, an online trading platform for aluminium products, and one of China's "big three" Internet portals. These case studies provide strong support for the theory of relationship-based commerce. The ability to overcome institutional deficiencies by building and maintaining strong relationships is a key success factor in each case.

Cultural factors have been commonly used to explain the distinctive patterns of IT application by organizations and societies around the world. In contrast, the evidence from these cases suggests that institutional factors, and institutional deficiencies in particular, are responsible for the networked and relationship-based forms of E-Commerce that are emerging in China. Implications for the research and practice of online commerce in developing countries and emerging market economies are considered.

1. Introduction

Enterprises across the People's Republic of China (China[25]) are no exception when it comes to electronic commerce. They are eager to modernize their businesses and realize the benefits of the Internet. Chinese managers have sought to learn from the experiences of dot-coms and Internet-enhanced businesses in the United States and other developed countries. Concurrently, *wise* men and women from the West, often touting themselves as e-business consultants, have been eager to advise Chinese clients. Unfortunately, the results of most efforts to develop E-Commerce in China have been disappointing.

The deflation of the dot-com bubble has exposed the general difficulties of profiting from online ventures. However, the lofty expectations associated with the emergence of the new economy reflected a widespread belief in the commercial potential of the Internet. This belief remains common among the leaders of the world's leading multinational corporations. As a result, the steady if less spectacular growth of E-Commerce continues in the United States and many other Western economies.

In contrast, an E-Commerce conference in Hong Kong that sought to apply the U.S. experience to China ended up being characterized as "the blind leading the bemused" (Pang, 2000). Experienced business people across mainland China tend to dismiss those touting the Internet as crazy and/or out of touch with reality. Meanwhile, a *Fortune 500* chief executive has asserted that E-Commerce in China was "hopeless" (Wong, 2000).

This pessimism in the Chinese context reflects the disappointing experiences of the vast majority who have ventured into E-Commerce (*China Online*, 1999; *Made for China*, 2000). Commercial Web sites in China, such as leading portal and content providers like sina.com, online book retailers like dandgang.com, and name-your-price auction sites like lalasho.com, have experienced difficulties similar to those of their American counterparts—Yahoo, Amazon, and Priceline, respectively. Even Chinese enterprises seeking to emulate the eBay business model, such as eachnet.com, experienced difficulties and were forced to branch into offline businesses to augment their sagging revenues.

The incremental convenience offered by online initiatives in China can also be underwhelming. For example, passengers wishing to use the online ticketing system of China Eastern Airlines had to live in Shanghai, buy a ticket originating from that city, use a credit/debit card from one particular bank, and then pick up their tickets at the airline's downtown office.

---

[25] The term China is used in this chapter to refer to the People's Republic of China. However, it excludes Hong Kong (despite Hong Kong's status as a sovereign component of the PRC, given the very different cultural, political and economic systems that prevail there).

1.1 Research Focus, Design and Method

Amidst this gloom and doom, a few Chinese enterprises are starting to benefit from electronic commerce. They have increased their market reach and/or share, improved their productivity and competitiveness, and have attractive rates of return on their E-Commerce investment. Longitudinal research begun in 1998 has examined these enterprises and considered the key factors behind their success. Consistent with the emerging nature of the phenomenon and our interest in understanding the "how" and the "why", a multiple case study design was used to study several of the early success stories in order to permit "replication" logic (Yin, 1994).

The unit of analysis was the E-Commerce initiative at the enterprise level. Each case was treated as an independent experiment in order to confirm or disconfirm emerging theoretical and conceptual insights. Data were collected using semi-structured interviews, observations, and assorted secondary sources. The primary data source was semi-structured interviews with individual informants. Key informants included managers and employees from the focal organizations and representatives from their major suppliers and customers. The sections that follow report on three of the pioneering success stories with E-Commerce in China—Mecox Lane, Li & Fung, and Asia Aluminum Holdings (see Table 10.1). We then outline the emerging success of one of the big three Internet portals in China (Sina, Sohu, and Netease) before considering the general pattern of online success that fits with the theory of relationship-based commerce (Martinsons, 2002).

Table 10.1

Profiles of the Emerging E-Commerce Success Stories in China

Company	Nature of Business	Year Established	Headquarters and Operational Base	Industry Position
Mecox Lane	Retailing	1996	Shanghai	Leader
Li & Fung	Supply chain management	1906	Hong Kong	Leader
Asia Aluminum Holdings	Aluminum extrusion and fabrication	1992	Hong Kong and Guangdong province	Leader
Sohu.com	Internet portal	1996	Beijing (incorporated in U.S.)	One of 3 leaders

2. From Mail Order to Online Retailing

Mecox Lane Holdings was established in 1996 as one of the first mail order businesses in China. Despite taking its corporate name from a street in the Long Island, New York resort town of Southampton, the operations of the company are based in Shanghai—the heart of the Chinese dragon. Mecox Lane achieved steady revenue growth by offering a wide product line and nurturing a market that had no prior experience with catalogue shopping. It has reduced many Chinese consumer doubts about the notoriously shady business of mail order shopping by developing relationships with government agencies including the national postal service. Customers can order, pay for, and pick up their Mecox Lane purchases at the local post office.

Mecox Lane has engendered customer loyalty by assuring product quality and building a strong brand image. It has also overcome the financial challenges posed by a conservative credit granting system and a relatively small credit card population. Its first coup was to obtain a commercial loan from a banking industry that has habitually favoured state enterprises over private firms, and made decisions based on political criteria as much as on financial ones. Secondly, Mecox Lane persuaded the banks to provide facilities that enabled mail order customers to make credit card purchases.

In April 2000, Mecox Lane went online to augment its catalogue sales business. This new marketing and sales channel was able to leverage the company's existing expertise, business processes and payment system. A full-page advertisement on the back page of Mecox Lane's mail order catalogue kick-started its presence on the World Wide Web. Its Web site exceeded a million registered hits and US$100,000 in online orders in its first month and revenues have continued to grow steadily over time.

It was not so easy for Mecox Lane to assure government officials that its new marketing and distribution channel was legitimate. As an early entrant into the B2C E-Commerce arena, the company went through a long and difficult process in order to get its Internet-based business model accepted by the state. Eventually, it convinced top officials of both the Shanghai Municipal Government and central government agencies that its online retailing venture was in the best interests of the local community and China overall. The value of cultivating relationships with the government became apparent when bureaucrats shut down several online retailers who asked to be forgiven after starting up rather than obtaining the assortment of permissions beforehand.

In contrast to the specialization and outsourcing philosophy that is common in the United States and other market economies (Markus, 2001), Mecox Lane handles almost every aspect of its business. In its early days, the firm contracted out the production of its mail order catalogue to an advertising company, but the results were disappointing. Now, it does its own production and is able to quickly modify its online catalogue. Mecox Lane routinely photographs a new product, uploads the results onto its Web site, and makes the product available for sale on the very same day that it is received from a supplier.

By the end of 2000, about one-third of Mecox Lane's revenue came from E-Commerce and it had reduced its marketing costs significantly. For example, customer surveys are now much less expensive because they can be conducted online. Although E-Commerce has cannibalized some of its mail order business, Mecox Lane doubled revenues in 1999, and again in 2000.

The company is now enjoying the economies of scale and benefits from cross-selling synergies. In the absence of a Better Business Bureau or other consumer protection agencies, Mecox Lane built up its business by gradually gaining the trust of mail order customers. Many of them followed the company online while new customers were attracted to Mecox Lane by word of mouth and advertising that has highlighted its *guanxi*[26], including friendly relationships with influential public agencies.

Along with another early entrant into online retailing, Beijing-based 8848.net, Mecox Lane pioneered cash-on-delivery services and worked with banking partners to enable credit card payment. Large and expensive items tend to be paid for with cash on delivery while smaller items are purchased with debit/credit cards and mailed to the customer.

Mecox Lane's choice of Shanghai was no coincidence. The metropolis is China's financial and trading hub. By early 2001, it had over 12 million credit/debit card holders, three million Internet users, a million homes with PCs, and a modern telecommunications infrastructure based on fibre optics (CNNIC, 2001). The Shanghai Commercial Value-Added Network was the first truly integrated E-Commerce support system in China. As a part of the Golden Card Project (Burn and Martinsons, 1999), this online network already provides reliable credit/debit card transactions processing and settlement. Hewlett-Packard and the Shanghai Municipal Government are now collaborating to further expand and upgrade the E-Commerce infrastructure.

Mecox Lane monitored the development of online retailing in both the U.S. and Hong Kong. It learned from the experiences of start-ups that initiated price wars against entrenched rivals. For example, AdMart in Hong Kong sold products below cost without achieving the economies of scale needed to sustain this low-cost provider strategy. In contrast, the price structure of Mecox Lane gives it a healthy profit margin. Moreover, the firm has differentiated itself by doing its own order fulfilment and providing superior customer service. In 1998, it made a large investment in a fully automated inventory control and fulfilment system. Mecox Lane subsequently developed an online transactions platform locally after products from U.S. vendors were deemed to be unsuited for functional as well as political reasons.

Mecox Lane provides specific benefits to both producers and consumers. Few Chinese manufacturers are capable of shipping their goods directly to end-users while

---

[26]   Guanxi is best translated as the network of relationships among various parties (often business) that co-operate together and support one another. For further details see Xin and Pearce (1996).

consumers have basic distrust of suppliers that they "do not know". Mecox Lane responded to the concerns about trust by hiring and training more than 100 front-end staff who now work in three shifts to provide around-the-clock customer service. Parallel teams of service representatives are responsible for telephone enquires and online communications.

Although well over 100 million Chinese now have a debit and/or credit card, the unlimited liability for credit card purchases and the lack of sales tax incentives for online purchases have limited their use. For example, Dell Computer received less than 7% of its orders from China online vs. about 40% of orders online in the rest of the world (Dell.com, 2001). Mecox Lane has mitigated some of the inherent drawbacks of E-Commerce with its focus on consumer trust, but its own surveys indicate that Chinese consumers remain very concerned about security issues (e.g., the reliability of encryption technologies) and legal issues (the enforcement of online commitments).

More than 20 million Chinese were regular Internet users in the year 2000 and about 80% of this group could be classified as having middle-class incomes (*China Online*, 2001; CNNIC, 2001). Mecox Lane has a database of 4 million Chinese customers who have demonstrated the willingness to purchase at least one product without seeing or touching it. Although this number can be expected to grow as Internet use increases across mainland China, it may be hindered by the lack of a rule-based system to govern commercial activities (Martinsons, 1999).

Could Mecox Lane become the Amazon.com of China? Although Mecox Lane has sold thousands of products to millions of customers, even its chief executive, Andrew Tsuei, has doubts. He says that "the Amazon.com business model will not work in China". He supports the theory of relationship-based commerce by suggesting that B2C transactions over the Internet will continue to reflect "E-Commerce with Chinese characteristics".

Unlike Amazon, Mecox Lane benefits from the fragmentation of the traditional retailing industry in China. Instead of large companies with nationwide coverage such as Barnes & Noble and Sears, the offline competitors for Mecox Lane tend to be poorly financed, single store operators that are the final link in an inefficient, multi-tier distribution chain. Foreign giants like Carrefour of France, Wal-Mart from the U.S., and Metro from Germany may emerge as the toughest competitors for Mecox Lane as they create national retailing networks in China (Kung, 2000).

Mecox Lane has part of a B2C E-Commerce market in China that totalled US$20 million in 2000, but this is meagre compared to the B2B arena where revenues amounted to US$80 million (Stout, 2001). The next couple of sections profile two of the emerging B2B successes in China. These cases suggest that leveraging and extending existing resources and relationships is not only a key to success for online retailing in China, but also for B2B marketspaces there.

3. A Well-Established Business Intermediary Goes Online

The highly fragmented nature of most industries in China has traditionally provided ample opportunities for intermediaries. Networks of small Chinese manufacturing enterprises have collaborated with each other and served as subcontractors to meet the needs of both domestic customers and large overseas buyers. Even after the rise of the Internet, long-established B2B intermediaries in China continue to thrive. In sharp contrast to the narrowing business scope of purchasing agents in the West (Milligan, 2000), intermediaries in China tend to manage the complexity and integrate the activities among dozens of small manufacturing enterprises.

With revenues that exceeded US$2 billion in 2000, Li & Fung Limited is the largest and most successful of these business intermediaries. The company was established in 1906 to export porcelain and silk from China. It is now headed by the grandsons of one founder and derives most of its revenues by supplying private-label goods, such as garments, fashion items, toys, and electronics, to clients like Abercrombie & Fitch, The Gap, and J. C. Penney. In recent years, Li & Fung has enhanced its market position by acquiring smaller industry rivals, including Inchcape in 1995 and Swire & Maclaine in 1999.

Li & Fung creates value by providing an array of supply chain management services. It takes on responsibility for sourcing, product design, shipment, warehousing, and letters of credit. As a trusted intermediary, it helps suppliers with marketing and buyers with procurement. One senior manager stated that a growing part of the company's value proposition comes from ensuring that customers "get the quality they desire at a price that they can afford". Li & Fung has invested huge resources over a long period of time to develop trust-based relationships with nearly 6,000 suppliers. Most of these suppliers are located in China, but a significant minority is based in other Asian countries.

As the foremost of the traditional B2B intermediaries in China, Li & Fung has used IT selectively to improve its business activities. Much of the focus has been on increasing the efficiency of its internal operations and managerial processes that span the small, customer-oriented units within the company. For example, an Intranet was introduced in 1995 to link the company's facilities and offices across the world. This internal IT application has significantly improved co-ordination and control.

In 1999, an E-Commerce subsidiary, StudioDirect, was established to extend its reach to a new market segment—smaller buyers in Western countries. Using the Internet, Li & Fung can aggregate its orders and achieve some economies of scale while continuing to offer some customization to each client. By December 2000, nearly 500 new buyers had been attracted to StudioDirect by the opportunity to obtain products such as shirts, shoes and home accessories with their own brand or label at an affordable price.

Li & Fung has also applied Internet technology to improve its communications with its long-standing customers. A number of secure extranet sites have been launched since 1997 to meet the needs of major customer groups. An increasing but still minor proportion of its sales orders are now taken over the Internet. Some parts of its product development and order fulfilment processes are also done online. According to one internal informant, these IT applications have reduced Li & Fung's unit costs by nearly 10% and shortened its average cycle time from order to delivery by 12–15%.

Efforts are now underway to set up online mechanisms for negotiating the terms of a contract and monitoring customer satisfaction. Another service being launched is eSO. This electronic stock offer application will provide Li & Fung's loyal suppliers with an alternative channel for disposing of their overstocked merchandise.

Despite the assorted applications that are part of its E-Commerce initiative, most of the value added by Li & Fung continues to stem from its decades-old relationships and intimate knowledge of manufacturing enterprises in China. Moreover, its ability to find the most appropriate combination of suppliers, and then to exert its influence on factory owners, reserving production capacity and assuring output quality, is largely independent of the information technology it uses.

Several new ventures in China, such as ChinaTradeWorld (www.chinatradeworld.com), MeetChina (www.meetchina.com) and Sparkice (www.sparkice.com), have relied almost exclusively on the Internet as a means to link buyers and sellers. Perhaps the new competitor with the highest profile is Alibaba (www.alibaba.com). Jack Ma, the founder of Alibaba, views his Web site as an electronic alternative to the trade fair. Business people can meet prospective partners in nearly 30 different virtual meeting rooms—each one specializing in a particular type of product or service. Although the site provides some information services, such as credit checks and shipping quotations, it does not have any online transaction capabilities. Therefore, Alibaba is an online meeting space rather than a marketspace.

Although the online upstarts in the China trade industry provide matchmaking services and, in some cases, online transaction capabilities, disintermediation by the Internet does not appear to be a serious threat to the business of Li & Fung. It would be very difficult for a dot-com venture to replicate the customized and personal services that this traditional intermediary provides.

U.S.-style B2B exchanges face some unique challenges in China. An online trading platform by itself has limited value to both Chinese suppliers and overseas buyers. Even with commodity products, payment and delivery issues must be addressed. With most goods and services, there are concerns about detailed requirements, quality and consistency, required down payments and credit terms, and how long it will take from ordering to delivery. Unfortunately, would-be electronic matchmakers in China have been hampered by an underdeveloped infrastructure and the limited amount of existing business services that are available to support online transactions. Nevertheless, at least one B2B marketspace (iMetal) in China has emerged as an early success story (see Hempel and Kwong, 2001).

4. Building a B2B Marketplace

i-Metal was founded in March 2000 by Asia Aluminum Holdings (AAH) (www.asiaalum.com), which at the height of the dot-com hyperbole had changed its name temporarily to Global Applied Technologies Holdings Limited. AAH operates China's largest aluminium extrusion and fabrication business and is listed on the Stock Exchange of Hong Kong. Pure aluminium is a commodity traded globally by metal producers, industrial users, traders, and financial industry players.

China is already a large producer and consumer of primary aluminium, with net imports equal to about 20% of its domestic consumption. Although the aluminium industry in China has tremendous growth potential, it differs from the West in terms of who competes and how they compete. According to government statistics, the top ten aluminium companies in China control only 30% of the domestic market compared to a 70% share of the U.S. market for its ten leading firms. The fragmented market has impeded supply chain efficiency and encouraged the presence of intermediaries to consolidate procurement.

Since domestic competition in the aluminium industry is based primarily on price rather than product quality, only a few aluminium firms in China meet the commonly-accepted export standards. These industry features favour the development of a B2B marketspace to expand market reach, to reduce marketing costs for suppliers, to reduce procurement costs for buyers, and to increase supply chain efficiency.

Nanhai, a city in Guangdong province near Hong Kong, is the hub for aluminium extrusion in China. It is home to eight of the top 10 producers and 380 of the top 1,000 aluminium extrusion firms registered in China. Since AAH has traditionally provided many services to the aluminium industry in Nanhai, its development of a marketspace may be seen as a logical extension of its existing business scope.

After initially considering the Internet as a potential new channel for marketing and sales, AAH decided to go beyond a simple on-line meeting place for potential suppliers or buyers, and to create an online trading platform that offers a wide range of bundled services. The platform handles trades of aluminium futures contracts on the Shanghai Futures Exchange, the physical metal using a B2B exchange model, and standardized finished products, including extrusions, using a B2B catalogue model. Buyers and sellers can also access real-time market data for both the Shanghai Futures Exchange and the London Metals Exchange. Most of the services are offered in three different scripts—traditional Chinese (used by the overseas Chinese), simplified Chinese (used in China), and English.

More significantly, Asia Aluminum has created business partnerships with domestic banks, logistic firms, and warehoused management firms to provide real-time electronic payment and settlement system, transportation services, and both bonded and non-bonded warehouse services. i-Metal is seeking to add value in this fragmented industry by establishing standards for both the technology platform and data exchange protocols. The development of value-added services is seen as an attractive means to create and sustain a competitive advantage.

Given the groundbreaking features of this marketspace, it is not surprising that i-Metal faces many design and implementation challenges. Similar to Mecox Lane, it had to prod the banking industry to enable the electronic transfer of funds. i-Metal.com worked with one of the four main Chinese banks to create a banking product called the Internet Commerce Debit Card. Even then integrating the Web site's trading platforms with the legacy systems of banks and brokerages on the Shanghai Futures Exchange required three months of intensive and often frustrating work. The arrangements necessary for this interfacing would have been impossible unless the founder of Asia Aluminum had a strong network of personal relationships in the local business community. The real-time funds transfer system now allows i-Metal users to quickly execute transactions in response to market changes.

The online trading platform for physical aluminium presented additional challenges. Unlike the well-developed rules and policies for trading futures on the Shanghai exchange, the online trading process for physical goods had to be developed from scratch. i-Metal ensured the integrity of the trading process by requiring both buyers *and* sellers to pay a 10% deposit. Here again, banks were coaxed to develop the capability to freeze funds in their accounts pending the conclusion of the transaction. i-Metal also helped to gain acceptance for collateral loans for commodities trading in China. In essence, i-Metal set out a conduct code for B2B marketspaces while encouraging Chinese banks to modernize. This should benefit businesses across China even as i-Metal seeks to capitalize on its first-mover advantage.

Security was another design issue for the i-Metal marketplace. It is addressing *external* threats by collaborating with China Telecom (a state recognized authority in this area) to introduce digital certification. However, in contrast to the West, Chinese firms are also very concerned with *internal* threats. In response, i-Metal has implemented multilevel security mechanisms into the marketspace so that only authorized staff can execute trades and transfer funds. i-Metal plans to incorporate bonded warehousing services into its marketspace design. This should increase market liquidity by enabling domestic buyers to purchase small quantities of metal that are physically located in Nanhai. With a bonded warehouse, the international suppliers of ingots need not pay duty until the metal is actually sold.

i-Metal is seeking to implement the system for trading of physical metals and to expand into the online trading of finished extrusions. Success with these efforts would make i-Metal a vertical B2B marketspace that provides an online trading platform for aluminium products ranging from primary materials (in the form of futures contracts and ingots) through finished extrusions. Consistent with a distinctive and deeply-embedded characteristic of Chinese business (Martinsons and Hempel, 1995), the marketspace design provides a single contact point for a range of industry services, but avoids higher-level business process integration.

i-Metal is envisioned to be a comprehensive services supplier akin to Li & Fung rather than just an online trading platform. The decision to include brokerage, logistics, and warehousing partners in the marketspace design results directly from this vision. i-Metal is seeking to capture the benefits of a network effect in order to become a dominant provider of online business services, first for the aluminium industry and perhaps later in other industries.

5. The Internet Portal

The ability to take advantage of both a network effect and economies of scale are evident in the rising fortunes of China's "big three" Internet portals: Netease (www.163.com), SINA (www.sina.com), and Sohu (www.sohu.com). By the end of 2002, each of these portals had achieved break-even financial results and was poised to enjoy the benefits of attractive profit margins on rapidly increasing revenues.

Sohu.com is an Internet portal that (in contrast to SINA) has maintained an exclusive focus on mainland China. The company was first established in 1996, as Internet Technologies China Inc. It launched its first website in January 1997 and spent much of the next year trying to gain customer acceptance while resolving assorted technical problems. In February 1998, the site was re-launched under the name Sohu, which in Chinese means "search fox".

Although Sohu.com's corporate offices are located in Beijing, the company is incorporated in the United States (Delaware) and its shares are listed on the Nasdaq National Market. Since regulations in China prohibit foreign companies from owning or operating telecommunications or Internet-based businesses, Sohu (and the other two leading portals in China) were forced to adopt an innovative corporate structure. The parent company loaned money to two of its top managers, Dr. Charles Zhang and Mr. Li Wei, so that they could set up a separate Internet content company. This company, Beijing Sohu, is owned entirely by Dr. Zhang and Mr. Li, both of whom are Chinese nationals.

The Chinese government subsequently approved Beijing Sohu's applications for licenses to develop Internet content, to provide information services, and to operate the Sohu.com website. Sohu.com provides both the information technology and the human resources that Beijing Sohu needs to carry on its online operations, and also pays a small monthly fee in return for receiving basic Internet services from the "domestic" company.

Sohu.com initially provided free access to e-mail services, assorted news content, and the world's first Chinese-language online search engine. Over time, the scope of services has been expanded steadily as a result of trial and error as well as customer feedback. By the end of 2002, the Sohu portal consisted of sophisticated Chinese-language Web navigation and search capabilities, 15 primary content channels, assorted Web-based communication and community services, and an extensive platform for E-Commerce and short messaging services.

Sohu.com's interest-specific content channels focus on popular topics such as news, business, entertainment, sports, and careers. Each channel encourages the development of personal relationships among those with a common interest in the topic. According to the annual survey conducted by Sinomonitor International, Sohu.com was the most visited Internet portal in China during both 2001 and 2002.

In an attempt to thwart the potential for direct competition from better-resourced international portals (such as Yahoo!), Sohu has sought to develop a lot of culturally-sensitive content that aims to meet the emerging needs of China's young and affluent urban population. This segment of the Chinese consumer market has grown up with the Internet, typically spending more time on this medium than watching television.

Sohu.com has three primary sources of revenue: advertising, communication services, and E-Commerce. Advertising accounted for about half of the company's revenues in 2002. The youth and affluence of Sohu's user base are appealing to domestic SMEs (small and medium enterprises) and the growing number of multinationals doing business in China. The active involvement of the CEO in Sohu's advertising sales efforts is a testament to the personal nature of business relationships in China. The chief executive, Dr. Zhang, regularly meets with the top managers of potential clients and gives public speeches that stress the value of online advertising. Advertising consists of banners, logos and links to their own Web sites, sponsorship of particular Sohu content areas, and bundled marketing services on the Sohu.net platform.

Sohu has recently established business relationships with both the (US) National Basketball Association (NBA) and its top Chinese player, Yao Ming. Sohu hosts, provides content for, and promotes the official NBA site in China. Basketball is one of the most popular sports in China, and the migration of top Chinese players to the NBA has spawned a growing interest in this league across the country. As a result, Sohu's basketball-related sites are expected to be an attractive option for advertisers seeking to reach a key segment of the Chinese consumer market.

The fastest growing revenue stream for Sohu.com and its peers has been subscription and messaging services. Short messaging services (SMS) have become very popular because Internet access in China using a mobile phone tends to be both more convenient and more affordable than using a personal computer. SMS enables users to send short messages (typically with a maximum of about 100 characters) across a variety of platforms to any other mobile phone user in the world at a much lower cost than speaking to them directly. The rapid development of these services can be attributed partly to the Chinese cultural preference for personal and informal forms of communication (see Martinsons and Westwood, 1997). It also reflects the power of the so-called network effect, whereby the value of the service to a given user increases exponentially with the total number of users.

Sohu.com has a universal registration system, so that someone who has established an e-mail account on the portal is automatically registered for its other services, such as instant messaging and bulletin boards. Significantly, Sohu has developed mu-tually-beneficial working relationships with the leading mobile phone service provid-ers in China. As a result, its registration system seamlessly includes every mobile phone user. These users can then access and download information, images, messages or games from the portal. Sohu.com receives micro-payments (a few cents or fen per access) from the mobile phone service providers when their customers use Sohu.com content. The emergence of both multimedia messaging services (MMS) and massively multiplayer online role-playing games (MMORG) offers new business opportunities for Sohu.com in partnership with mobile phone service providers and online game developers, respectively. For example, one Chinese market research firm, CCID Con-sulting, projects that the customers for online game playing in China will grow from a base of about 10 million in 2002 to triple that number by 2005. Another report, pub-lished by DFC Intelligence (WWW.DFCINT.COM) suggests that the direct revenues from online games in China will exceed US$500 million in 2005.

With an increasingly popular brand image and relatively stable top management team, Sohu.com is also steadily gaining the trust of Chinese consumers. This is re-flected in increasing numbers of online purchases through its portal. Sohu.com sells a wide range of products and services (some of which compete directly with Mecox Lane) on its E-Commerce site and offers online securities trading through Sohus-tock.com, which is the result of a joint venture with a leading Chinese brokerage firm, Guolian Securities.

6. E-Commerce Success in China

Reports on E-Commerce development in the United States and other Western coun-tries tend to focus on reduced overhead costs and increased convenience as a result of more transparent information and pressures for changes in industry structure such as market disintermediation and supply chain integration. The fundamental premise is that faceless market transactions enabled by IT will improve economic efficiency and the quality of life.

Businesses across China have applied IT to improve their internal co-ordination and control (Martinsons and Westwood, 1997). However, online business initiatives using Internet-based tools and other IT applications have tended to be less common and less successful. The rare cases of E-Commerce success in China tend to resemble electronic networks, with co-operation among complementary product/service pro-viders, more than electronic markets (Malone *et al.*, 1987). The emerging success sto-ries have responded to institutional deficiencies by developing or leveraging both per-sonal and business relationships based upon trust and mutual benefit. These connec-tions have enabled them to overcome regulatory, financial and logistics issues (see Table 10.2).

In October 1978, in a statement to the Party Congress, Deng Xiaoping insisted that the economic reforms in China represented the implementation of socialism with Chinese characteristics. A similar adaptation of a foreign concept to the circumstances in the People's Republic of China is now evident. Mecox Lane, Li & Fung, i-Metal, and Sohu.com may be seen as exemplars of E-Commerce with Chinese characteristics. These emerging success stories have chosen to focus on providing value-added intermediary services. They are not *disintermediating* the market by displacing existing middlemen. Instead, they are using the Internet to enhance user experiences (and in some cases, enhance the efficiency of the market) by developing entirely new or at least significantly improved services.

Table 10.2
Relationships—The Key to E-Commerce Success in China

Company	Relationships	Benefits for E-Commerce
Mecox Lane	Government	Authorization for online retailing
	Banks	(Credit card) Payments
	Post office	Product distribution
Li & Fung	Thousands of suppliers	Ability to attract online customers
	Large customer base	Online segmentation/customization
Asia Aluminum Holdings	Government	Authorization for online trading
	Banks	Credit and collateral services
	Delivery service providers	Transport of traded aluminium
	Warehouse owners	Warehousing of traded aluminium
	Telecom services provider	Online security for trading platform
Sohu.com	Government	Authorization for online services
	Hundreds of business allies	Content and advertising
	Telecom services providers	Messaging services (SMS/MMS)
	Guolian Securities	Sohustock.com (online broker)
	Large customer base	Online segmentation/customization

Mecox Lane, i-Metal and Sohu.com are all working with national-level partners to develop support services for their marketspaces. As first (or early) movers in designing these services, these enterprises may serve as reference models for future national standards or policy reforms. For example, Mecox Lane and i-Metal have deliberately designed support services that follow international standards in order to prepare the marketplace for China's entry into the World Trade Organization (WTO). Significantly, their business partners have not been required to formalize or redesign their business processes or to share management information. Even these precursors to supply chain integration remain the exception rather than the norm in Chinese economies (see Martinsons and Hempel, 1998).

E-Commerce activities in China face a common uncertainty: regulation and interference by the state and its officials. As noted in the case of Sohu.com, many online business activities, and particularly those involving foreign partners and business networks, are tightly monitored and controlled by government agencies. For example, the online brokers of securities and commodities in China are prohibited from discounting their commission rates. More generally, the overlapping authority of state agencies and the arbitrary enforcement of enacted laws have shaped the development of E-Commerce in China.

There is little doubt that the Communist Party *needs* China to flourish economically. Anything less would imperil social stability, undermine the Party's legitimacy, and consequently threaten its monopoly on political power. Communist Party leaders also appear to recognize that modern technologies, including the Internet, are essential for economic progress (Kennedy, 2000). The Government Online Project represents an effort to lead by example and it has improved Chinese government transparency. However, the Web sites of public ministries and agencies linked to www.gov.cn tend to fulfil a broadcast function, providing basic information and touting recent achievements rather than offering online services to Chinese citizens.

Some powerful and privileged interests in China continue to impede the country's transition to E-Commerce. They worry about the social and political consequences of such a transition. The Internet has already undermined the Chinese Communist Party's monopoly on "public" information. For example, Internet portals such as Sohu and Sina played an instrumental role in alerting Chinese people to the outbreak of severe acute respiratory syndrome (SARS) in early 2003 even as reports on this epidemic were notably absent in official news sources.

The Chinese government continues to expend vast resources to suppress online dissent and to regulate what its people see and do on the Web. Excessive state oversight has raised concerns about the confidentiality of online transactions while ambiguous laws make it difficult to authenticate users or to confirm transactions that would avoid repudiation. Top managers from online enterprises such as Mecox Lane and Sohu.com have expressed their concerns about this issue in various public forums.

Improvements in the environment for E-Commerce may be expected as China continues its domestic reforms and satisfies its requirements as a member of the WTO. However, a key question can only be answered over time: how (and how quickly) will Chinese business and society change as the faceless world of E-Commerce penetrates its economy?

7. Conclusions

The trials and tribulations of dot-com firms and online ventures by traditional firms all over the world demonstrate the difficulty of doing business on the Internet, and especially doing it profitably. Nevertheless, the development of E-Commerce in the U.S. (and other modern market-based economies) has been stimulated by the presence of convenient and inexpensive Internet access, well-developed on-line payment and physical transportation systems, and many affluent consumers. Perhaps most importantly, these market-based economies have comprehensive, impartial and enforceable rules to govern market transactions.

Traditional commerce, both at the retail level and between companies, remains comparatively inefficient in mainland China. However, the environment for E-Commerce is also very challenging. The personal nature of business relationships, the informality of information, and the lack of separation between political and economic actors are among the factors inhibiting and shaping the development of B2B E-Commerce in China. Meanwhile, Chinese consumers, who have no tax incentives to shop online and little legal recourse if the transaction goes wrong, have been reluctant to abandon traditional marketplaces where they can enjoy a social experience and engage in time-honoured rituals such as haggling or asking for special deals.

The success of the leading Internet portals in China has been based largely on products and services that foster social interaction and personal relationships. Short messaging services sent by one mobile phone user are helpful to cultivate personal relationships while massively multiplayer online games enable collaboration among large groups of people.

The business environment in China, and specifically an underdeveloped set of business rules (Martinsons, 1999), has encouraged distinctive forms of E-Commerce with Chinese characteristics. Each of the profiled success stories used the Internet to create or extend a competitive advantage by building upon its core competencies. To be specific: Mecox Lane is leveraging the marketing expertise and order fulfilment capabilities from its successful mail order business model; Li & Fung has gone online to reinforce its knowledge of and relationships with thousands of suppliers across China; and Asia Aluminum Holdings has used its market leadership position to modernize the aluminium industry's supply chain with its i-Metal venture. Meanwhile, Sohu.com is seeking to rapidly increase its revenues while maintaining its attractive profit margins by providing innovative services that will appeal to an expanding user base. Given the enduring nature of Chinese relationships and the fragmentation in their industries, each of the profiled ventures should enjoy a significant early-mover advantage. It remains to be seen whether they can capitalize on that advantage and sustain a competitive edge.

In mainland China, E-Commerce may not only enhance efficiencies at the firm or supply chain level, but also serve as a catalyst for hastening the development of the fundamental rules and infrastructure needed for market-based commerce. Although the Chinese government aims to promote IT-enabled economic development, its Internet policies have the potential to inhibit business innovation. It remains to be seen how electronic commerce will develop in China and what role it will play in the economic development of the world's most populous country.

Acknowledgements

This paper is derived from case material presented by the author at the Academy of Management meetings in Toronto (August 2000) and Denver (August 2002), at the University of Melbourne (December 2000), Peking University (July 2002), and the Stockholm School of Economics (July 2002). The author gratefully acknowledges the insights and comments provided by David Chui, Robert Davison, Paul Hempel, Yi King Kwong, Alan Li, Shaomin Li, Shuhe Li, Lynne Markus, Aelita Brivins Martinsons, Doug Vogel, and Zhou Nan. He also wishes to thank stakeholders in the profiled enterprises for their service as key informants.

References

Burn, J. M and Martinsons, M. G. (1999) Information Technology Production and Application in Hong Kong and China: Progress, Policies and Prospects, in: Tan, F. B., Corbett, S. and Wong, Y. Y. (eds.). *Information Technology Diffusion in the Asia Pacific*. Hershey: Idea Group Publishing, 7–35.

CNNIC—China National Network Information Center (2000-2002) Numerous publications, press releases and online postings. www.cnnic.com.cn

China Online (1999) E-Commerce in China: Horse-drawn Buggies on the Information Highway, 29 December, extracted from a report in *Zhongguo Shang Buo* (*China Commercial News*), www.chinaonline.com/issues/earchive.

China Online (2001) Dungeons and the Dragon: E-Commerce in China, 3 January, www.chinaonline.com/issues/earchive

Hempel, P. S. and Kwong, Y. K. (2001) B2B E-Commerce in Emerging Economies, *Academy of Management*, 6 August, Washington, DC

Kennedy, G. (2000) E-Commerce: The Taming of the Internet in China, *China Business Review*; **27**, 4, 34–39.

Kung, L. (2000) Rise E-distribution: E-Commerce with Chinese Characteristics, *MFC Internet Update*, 25 August, www.mfcinsight.com/article/000825/kung.htm

Made for China (2000) Insight Newsletter on B2B E-Commerce, 15 July, www.mfcinsight.com/article/000715/oped2.htm.

Malone, T., Yates, J. and Benjamin, R. (1987) Electronic Markets and Electronic Hierarchies, *Communications of the ACM*, **30**, 6, 484–497.

Markus, M. L. (2001) Paradigm Shifts—E-business and Business/Systems Integration, *Communications of the AIS*, **4**, 10, http://cais.isworld.org

Martinsons, M. G. (1999) *Relationships not Rules for Business in China.* Vancouver: Pacific Rim Institute for Studies of Management.

Martinsons, M. G. (2002) *A Theory of Relationship-Based E-Commerce.* Vancouver: Pacific Rim Institute for Studies of Management.

Martinsons, M. G. and Hempel, P. S. (1995) Chinese Management Systems: Historical and Cross-cultural Perspectives, *Journal of Management Systems*, **7,** 1, 1–11.

Martinsons, M. G. and Hempel, P. S. (1998) Chinese Business Process Re-engineering, *International Journal of Information Management*, **18**, 6, 393–407.

Martinsons, M. G. and Westwood, R. I. (1997) Management Information Systems in the Chinese Business Culture: An Explanatory Theory, *Information and Management*, **32**, 5, 215–228.

Milligan, B. (2000) Third-party Logistics Expands Reach, but Falls Behind on Internet, *Purchasing*; **128**, 3, 81–85.

Pang, L. (2000) Here be Dragons—The Perils of Mapping E-Business, *Gorilla Asia*, gorillasia.com/features/rr-dragons30oct00pang.htm

Stout, K. L. (2001) B2B Dominates China E-Commerce, *CNN News Online*, 22 February, www.cnn.com/2001/WORLD/asiapcf/east/02/22/hk.chinab2b.htm

Wong, K. C. (2000) Multinationals Frustrated by E-Commerce, *China Business Online*, 17 November, www.chinabusiness.com/articles/November/Wong2.htm

Xin, K. R. and Pearce, J. L. (1996) Guanxi: Connections as Substitutes for Formal Institutional Support, *Academy of Management Journal*, **39**, 6, 1641–1658.

Yin, R. (1994) *Case Study Research: Design and Methods*, (2nd ed.) Thousand Oaks: Sage.

About the Author

Maris G. Martinsons is a professor of Management at the City University of Hong Kong and Research Director of the Pacific Rim Institute for Studies of Management. He holds engineering and MBA degrees from the University of Toronto and a PhD from the University of Warwick.

Maris has served as an editor for *IEEE Transactions on Engineering Management*, the *Journal of Applied Management Studies*, the *Journal of Information Technology Management*, and the *Journal of Management Systems*. His professional interests focus on the strategic and cross-cultural issues associated with the management of information resources and organizational change.

Maris is widely acknowledged to be a leading authority on business and management in Chinese societies.

11. E-Procurement by the Brazilian Government

Luiz Antonio Joia

Brazilian School of Public and Business Administration
Getulio Vargas Foundation
and
Rio de Janeiro State University
luizjoia@fgv.br

Fuad Zamot

Brazilian School of Public and Business Administration
Getulio Vargas Foundation
fuad_zamot@uol.com.br

Abstract

The purpose of this paper is to evaluate the use of Internet Technology in the process of e-procurement currently employed by the Brazilian Federal Government. The focus of this evaluation is on the Internet-Based Reverse Auction system widely employed by the Federal Administration to purchase goods and services, from the standpoint of efficiency, efficacy and accountability. An explanatory single case study methodology addressing web-based purchase of pharmaceutical products by the Ministry of Social Security is used to answer the research questions raised in this paper. Technical, human and structural obstacles, as well as legal constraints that must be overcome, are also addressed. Conclusions detailing some findings and the advantages and disadvantages of this new *modus operandi* are also presented in this work.

1. Introduction

At this point in time, it is still difficult to grasp fully the meaning, opportunities and limitations of the "electronic government" concept (Prins, 2001). This does not mean that important steps have not been set in motion to achieve an information-based government. To date, governments have widely recognized the potential of new Information and Communication Technologies (ICT) to bring about fundamental renewal, not only in their operation but also in their position *vis-à-vis* other organizations, societal groups, or individuals (Fountain, 2001).

The objective of this study is to identify, in a general way, how electronic government can contribute to constructing a more agile, flexible and responsible State for its citizens. More specifically, this study evaluates Internet-based reverse auctions staged by the Brazilian Federal Government, analyzing the Electronic *Pregão*[27] no.12/00—conducted by the Ministry of Social Security (MPAS) for the purpose of acquiring pharmaceutical products—from the standpoint of efficiency, efficacy, and accountability.

Hence, the main questions this research purports to answer are:

(i)   How has the procurement process of the Brazilian Government using the Electronic *Pregão* improved in terms of efficiency?

(ii)   How has the procurement process of the Brazilian Government using the Electronic *Pregão* improved in terms of efficacy?

(iii)   How has the procurement process of the Brazilian Government using the Electronic *Pregão* improved in terms of accountability?

2. Contextualization of the Problem

According to McHale (1972), we live in an era of critical transformation, revolution and change. The rethinking of organizations and their productive processes has been triggered by the increasing development of Information and Communication Technology (Joia, 1994).

This was precisely the focus of a survey carried out by The Economist (2000), which stated that after electronic commerce and electronic business, the next Internet revolution would be electronic government. By electronic government, in its simplest form, we should understand it to involve the application of information and communication technology (ICT) to the process of public administration. There are advantages that can be clearly verified both by government and by the citizen. For govern-

---

[27]   Pregão is the Portuguese expression used in Brazil for "Reverse Auction staged by the Federal Government"

ment, even though some data is lacking, savings of approximately 20% are estimated if the private sector patterns are accomplished. According to the survey, as per the U.S. budget, savings could be in the order of US$110 billion per year.

In Brazil, the Federal Government's requirement is US$4 billion. The changes are also significant for the citizen. Besides being able to access a 24-hour-a-day service, 7 days a week, with no lines and none of the usual bureaucracy, the technological and management innovation not only changes the way the government provides its services but also allows for re-structuring the role of the state in relation to society with respect to transparency, democratic procedures and citizenship participation, in accordance with a document published by the Organization for Economic Co-operation and Development (OECD, 1998). Integrated e-government makes it feasible and desirable to build a new architecture for "seamless government", no longer consisting of a range of "stove pipe" organizations but of networks connecting "one-stop" front offices to the back offices of service providers (Leitner, 2003, p. 33). The same is true for corporations, as they have created real value for changing their processes, creating business models and redefining structures in their industries (Maira and Taylor, 1999).

The public administration's view can be judged from the opinion of the team co-ordinator of the Brazilian Electronic Government's site (Osório, 2001), who argues that the State's relationship *vis-à-vis* society becomes clearly defined in two segments as a result, namely the Citizen and Society segments. In the Citizen segment, through government, the State pursues the objectives, as perceived by citizens, to be attained in the form of public policies such as those relating to monetary, fiscal, sanitary, social security, transportation development and consumer rights aspects, among the many areas in which it must interact. In the Society segment, mainly through public policies adopted by the government, the State executes or creates conditions for the public administration to ensure that society may obtain the services desired by various groups: citizens, corporations, other governments (sub-national or supranational) and insiders—by the administration itself. In both cases, the intensive use of ICT during the second half of the 90s has allowed for an effective increase in efficacy in the former segment, and efficiency in the latter.

As already indicated, the second segment, i.e., the Society segment, includes the provision of services to society, which can be done via the government/citizen channel (G2C), the government/corporate and financial market channel (G2B/G2I), as well as the government/government (G2G) interface at different levels and spheres. The scope of this research is the study of the digital channel established for the purchase of goods and services by the Government (G2B).

Starting in 1993, several Ministries began using the Internet to transmit information of interest, mainly the Ministry of Finance and the Ministry of Science and Technology that co-ordinated the Brazilian academic Internet, or RNP. From 1994 onwards, the impact of the Internet provoked the acceleration of the review process of the model that had previously been adopted. Several local companies succeeded in implementing a marked process of technological and even administrative modernization, for it was at this point that the renovation process of the State apparatus was initiated. In 1995, the quantity of federal government sites increased exponentially on the Web. In more recent years, the variety of sites has increased not only in the federal area but also in the state sector with a focus on systematisation of information in federal government by the Ministry of Planning through the site located at: www.redegoverno.gov.br.

An important step towards instituting more modern and efficient tools as well as transparency in governmental procedures was the creation of "Net Purchase" (net shopping), an on-line system which permits access to notices with invitation to bid, price determination and public tenders conducted by federal and all other levels of administration. In addition to all the initiatives outlined above, the Electronic Auction (*Pregão Eletrônico)* came on stream at the end of 2000, in a clear attempt to reduce costs incurred by the State when staging public tenders for contracting products and services from private companies.

The Planning Ministry is considering adopting the *Pregão* for all acquisitions whenever required, although it will be used exclusively to contract services and buy goods, the performance and quality of which can either be objectively appraised by "government issue" through standard market specifications, or can be offered by several suppliers and easily compared, in order to permit decisions based on the lowest price. It does not apply, for example, to engineering works and services, which will continue to abide by the rules of public tender law N. 8666/93, as will be presented later in this paper.

2.1 Traditional Process of Government Public Tenders in Brazil

Machado (2001) analyses the statute on public tenders in Brazil and its evolution over time, clearly illustrating the trajectory of Brazilian legislation concerning public sector purchases. He singles out the law of 29 August 1928 as a landmark in legislation and states that some references are still in effect such as, for example, the definitions of the types of competition, the publishing deadline of the invitation to bid and other examples. He points out that during the period in which the country was under military rule, law number 4401/64—which consolidated the public tender concept— was enacted, in addition to Decree-Law 200/64, which constituted a key aspect of the administrative renovation proposed by the government. This renovation aimed at creating an agile, decentralized and de-concentrated state, in order to implement the agility of the private sector in the public sector. However, it is common knowledge that there was a great deal of abuse.

The next step was Decree-Law 2300/86, which remained in force until the drafting of the 1988 Constitution. Law N. 8666/93 was elaborated for regulating art. 37, paragraph XXI, of the Federal Constitution, and it was passed within a singular historical context marked by public concern in relation to accusations of corruption. This gave rise to a wealth of detail in relation to procedures in an attempt to wrap up the subject conclusively, leaving little room for self-regulation or adaptation of procedures to specific circumstances. Still, according to Machado (2001), the text of this same law stipulates the following form of tender procedure for purchasing and contracting.

(i)   Public Bid—Those interested shall fulfil the qualification prerequisites required by government edict; this applies to purchases in excess of R$650,000.00[28]—or R$1,500,000.00 (in the case of engineering works and services)

(ii)   Price Determination—This allows for previously registered suppliers bidding to supply amounts up to the limit mentioned above.

(iii) Invitation to Bid—This form involves the selection and invitation to bid of a minimum of three suppliers in the same region for the desired product or service. It can also be adopted for purchases over R$80,000.00 or R$150,000.00 (in the case of engineering works and services)

(iv)   Public Competitive Examination—This form is adopted for choosing technical, scientific or artistic work though a prize or earnings, according to the criteria defined in the government issue.

(v)   Auction—This is the method adopted for sale of discarded goods and impounded or pawned products through bids.

For judgment criteria, the law stipulates that the alternative with the lowest price, best technical aspects or lowest price combined with best technical aspects shall be the winner.

2.2 The Electronic Pregão in Brazil

It was the recent Provisional Measure N. 2-026/2000 of 5 April 2000, regulated by Decree 3555/00 of 8 August 2000, which established the *Pregão* as being the chosen procedure for the acquisition of basic goods and services up to any amount throughout the Federal Union, including the state legislative and judiciary. Purchases that must be conducted on the combined basis of technical aspects and price criteria will be excluded from this option. The Electronic *Pregão* in Brazil was regulated by the decree of 21 December 2000, which controls the utilization of Information and Communication Technology.

According to Braga (2001), there are two mechanisms under scrutiny that characterize the Electronic *Pregão* as a tool for speeding up services and eliminating bureaucracy, namely the definition of the lowest price and the straightforward qualification of a winning contestant.

The Electronic *Pregão* can be adopted for the same types of purchase and contracting operations conducted by competitive bid. It can be conducted in public session through a system that uses communications via the Internet. Furthermore, the contestants may place electronic bids making use of encryption resources, thereby guaranteeing information security. Consumer goods and services may be acquired via the Electronic *Pregão,* once the patterns of performance and quality are clearly defined by an invitation to bid using specifications in current use on the market.

---

[28]   US$1 equals nearly R$ 3 (circa June 2003)—R$ (real) is the Brazilian currency

Consumer goods include:

Mineral water, fuel and lubricants, gas, food, hospital, medical and laboratory material; pharmaceutical products, drugs, cleaning and maintenance material, oxygen and uniforms, furniture and equipment in general, excluding information technology goods, motor cars and vehicles in general, PCs and portable microcomputer (laptops, notebooks), video monitors and printers.

Services include:

Administrative support (keyboarding and maintenance), subscriptions, hospital, medical and dental assistance, auxiliary activities, uniform confection, events, film shooting, gas supply, graphics, hotel services, gardening, laundromat, cleaning and maintenance, rental, removal and maintenance of movable and non-movable goods, microfilm, health insurance, translation, data telecommunications, image and voice, fixed and movable telephone services, transportation, meal vouchers, watchmen and security, electric power supply, maritime support services, training and further qualification services.

3. Research Methodology

A single case study methodology will be used in order to analyze the efficiency, efficacy and accountability of the Electronic *Pregão* developed by the Brazilian Government. The main focus is on the procurement process elaborated by the Ministry of Social Security, hereinafter called the MPAS, to purchase pharmaceutical products from several potential suppliers.

Case studies are particularly suitable for answering "how" and "why" questions, and are ideal for generating and building theory in an area where little data or theory exists (Yin, 1994). It also enables the researcher to use "controlled opportunism" to respond flexibly to new discoveries made while collecting new data (Eisenhardt, 1994). Embedded single case research methodology (Yin, 1994) was used in this paper, as multiple units of analysis were taken into account and measured. Yin's tactics (construct validity; internal validity; external validity; and reliability) were carefully considered in this research.

In particular, construct validity was dealt with in the study through the use of multiple sources of evidence—as several bids relating to different pharmaceutical products were examined and collected—, the establishment of a sequence of evidence, and having the members of the group review the draft case study report. Internal and external validity in the findings was also taken into account, mainly by applying statistical analysis and replication logic respectively. Finally, the reliability of the results was ratified using a case study protocol and developing a case study database.

An explanatory approach was applied in the case study. Explanatory case studies are useful for assessing how and why an intervention is working. The methodology verifies whether problems and modifications are needed in the intervention, and try to explain the causal effects found. Different sites are needed in order to develop a comparative analysis. (Morra and Friedlander, 1999).

4. Theoretical Reference

4.1 E-Procurement

According to Kalakota *et al.* (1999), there is great confusion between purchasing and procurement. Purchasing refers solely to the process of buying goods and services, whereas procurement is a macro-process involving not only the process of purchasing of services and goods but also the logistics, storage, reception, inspection and monitoring processes. E-procurement might be interpreted as the procurement macro-process developed with the help of Internet technology. According to Neef (2001), there are three types of e-procurement:

- buy-side desktop requisitioning—the employees, themselves, through their desktops and using the corporate intranet and its link with the Internet, undertake on-line purchase, complying with the company's buying routines and procedures;
- buy-side centralized procurement management—purchasing managers (for instance), on behalf of the company, control the whole procurement process, analyzing transaction data and undertaking the management of the suppliers;
- sell-side applications—solutions developed by potential suppliers to help them negotiate their products and services on the web.

The E-Procurement solution adopted by the Brazilian Federal Government is based on the buy-side centralized procurement management model. Hence, this will be the type of e-procurement analyzed in this paper, the *modus operandi* of which is presented in the "Reverse Auction on the Web" section below.

4.2 Performance Measurement

According to Ballantine and Cunningham (1999), increasing recognition of the need to monitor multiple dimensions of performance has led to the development of a substantial body of Performance Measurement literature (see, for example, Brignall *et al.*, 1992 and Fitzgerald *et al.*, 1991, to name only two). Among the earlier contributors to the literature, Checkland *et al.* (1990) conceptualized Performance Measurement by using the concept of a system and the measures necessary for it to remain stable over time. Their research led to the recognition of three levels of performance which, they argue, should be used to monitor a system's performance (see also Checkland, 1981):

- *Effectiveness*: Is the right thing being done?
- *Efficacy*: Does the means work?
- *Efficiency*: Is resource usage minimal?

Roebeke (1990) largely concurs, recognizing the need to monitor effectiveness, efficacy and efficiency. He suggests that the three criteria constitute a hierarchy, within which measures of effectiveness are of more importance than measures of efficacy, which in turn are more important than measures of efficiency.

According to Morkate (1999), something is efficacious if it succeeds or does what it should do. He defines efficacy as a way to establish an objective to be met, which must include the quality of what is proposed. Further, he states that this objective must delimit a time at which one hopes to generate a determined effect on the product. For this to occur, an initiative becomes efficacious if one meets the expected objectives at a programmed time with the expected quality.

Accountability is an expression which originated from the English language to designate mechanisms of charging, billing, or, in other words, procedures of public responsibility of the governors (Campos, 1990). To Campos (1990), this concept may be understood as a question of democracy. The more advanced the democratic stage, the greater the interest in democracy. And government accountability tends to follow the advance of democratic values such as equality, human dignity, participation, transparency and responsibility.

4.3 Reverse Auctions on the Web

Auctions are an age-old mercantile practice, namely the public sale of objects to whomsoever places the highest bid. For Turban (2000), auctions offer opportunities to buyers and sellers that are not available using traditional means; however, whatever the type, they have many limitations. When they are conducted on the Internet, they allow access to a greater number of both individual and corporate participants at a lower cost. According to Blackmor's prediction, this market is expected to hit the US$2.7 trillion mark in 2004.

Turban (2000) further shows that electronic auctions have been around for many years like the one for pigs in Taiwan and Singapore, automobiles in Japan and the flower auction in Holland which was already using information technology resources back in 1995 (Kambil and van Heck, 1996). It was during the course of that same year that auctions began being conducted on the Internet.

Electronic auctions are broadly similar to traditional auctions except that they are conducted with the use of a new tool—the computer. On the Internet, host sites function as intermediaries offering services to anyone wishing to sell their merchandise and, for those seeking to buy, it's the ideal way of placing their bids.

There is no lack of information. It is available on-line or can be obtained by e-mail for items of greater value. Offers can be submitted via e-mail or by filling out electronic forms. Those that remain for longer periods are shown on the site. However, the names of bidders are kept secret from the other participants. It has been detected that some sites insist on codes of conduct that participants must abide by to conduct their transactions, for instance in order to prevent one bidder using various identities from placing multiple bids.

According to Wurman *et al.* (2001) there are several ways of classifying auctions. They emphasize the U.S. model, which is used when sellers offer many identical items simultaneously. In this category, offers are increasing exponentially and products are sold to the highest bidders. In the Dutch auction, generally open to the public and patented by Bid Com International of Ontario, Canada, the prices drop until buyers make an offer.

In Brazil, the Federal Government adopted the Reverse Auction based on the Dutch model for acquisition of goods and services, whereby the Government defines a base price, a maximum duration for the auction is established and companies place their bids via the web until the auction is finalized. As responsiveness is demanded by the society from the Public Administration, the Reverse Auction based on the Dutch model streamlines the procurement process undertaken by the government. Furthermore, this type of auction also complies with the legal requirements adopted by the Brazilian Government to deploy this new *modus operandi*. From the legal standpoint, the Brazilian Federal Government has decided to call this new form of public tender *Pregão*.

5. Case Study Analysis

The case study analyzed in this paper is that of *Pregão* N. 21/2001 conducted by the Co-ordination of General Services of MPAS, in line with the terms of the Electronic Auction Paper available on 18 August 2001 at the following address: www.comprasnet.gov.br/pregão.asp and opened at 3:00 p.m. on 9 August 2001, in accordance with legal precepts for purchase of pharmaceutical products, in order to supply the needs of the Ministry's Medical Assistance Pool.

As previously stated, the auction was opened at 3:00 p.m. on 9 August 2001 and suspended at 7:15 p.m. of the same day. It was re-opened at 8:00 a.m. on 10 August 2001 and closed definitively at 9:25 a.m. on 10 August 2001, with a total duration of five hours and forty minutes.

5.1 Data Collection and Analysis

- The *Pregão* included 164 items to be auctioned, 34 of which were not adjudicated.
- The reference value offered by the Ministry amounted to R$1,427,894.40 relating to 130 items auctioned.
- The total acquisition value relating to the 130 adjudicated items amounted to R$1,004,952.95.
- The 34 items cancelled because of lack of offer had been allocated R$143,030.15 (reference value).
- There is no record of appeals as the *Pregão* was closed at 9:25 a.m. on 10 August 2001

- The *Pregoeiro* (The Auctioneer in the Electronic *Pregão*) excluded some proposals either due to discrepancy or impracticability.

In Exhibit 11.1 below we can follow how, over the course of time, the negotiation for a specific item—in this case, item 2 (30 units with 120ml of paediatric acebrofilina coughmedicine)—occurred in the auction.

Exhibit 11.1
Evolution of Bids Placed, over the Course of Time (referred to in item 2)

Supplier	Bid Value R$	Date	Time
Company B	300.00	09/08/2001	15:02
Company A	299.70	09/08/2001	15:02
Company B	299.60	09/08/2001	15:08
Company A	299.50	09/08/2001	15:10
Company B	299.40	09/08/2001	15:12
Company B	299.40	09/08/2001	15:12
Company A	299.20	09/08/2001	15:15

As seen above, the variation between the total budget allocated for acquisition and the total budget spent, after discount, for 134 acquired items was of -29.62%, as follows:

- Total allocated: R$1,427,894
- Total spent: R$1,004.952

Therefore, the difference in its favour between the reference price used by the Ministry and the product acquisition value attained the impressive figure of nearly 30%. After careful analysis of the bids placed for all the items auctioned (128 observations), applying statistical tests for internal validity of the case study (Yin, 1994) and working with LN (bid)[29], in order to verify whether there is any statistical correlation between the number of bids for a product and the discount obtained by the Federal Government in the purchase of this specific product, it was seen that:

- Simple regression gave us $R^2=0.33$; which for a simple regression is significant ($\rho \sim 0.6$). Hence, there was a positive correlation between the number of bids and the price discount (as a percentile) on the items auctioned, i.e., the greater the number of bids, the lower the price paid by the Government for the corresponding item auctioned.
- The adjusted regression found was: Discount=7.5635 + 11.0856. LN (bid).

---

[29]   Neperian logarithm of the number of bids as a reference variable

- For each variation unit in LN (bid), we found an average discount of 1.0856% in the item auctioned.

6. Findings

6.1 Performance Evaluation

As stated above (Prins, 2001), at this early stage, it is still difficult to grasp the meaning, opportunities and limitations of the "electronic government" concept fully. However, one major conclusion can be drawn, namely that there is a pressing need to integrate e-government and E-Commerce (Kubicek and Hagen, 2001), and e-procurement is the ideal link to make such integration feasible.

On the basis of what has been outlined above and answering the research questions, it is possible to conclude that:

As to efficiency, it has been demonstrated that through the Electronic *Pregão* model it is possible to achieve:

- A reduction of nearly 30% in product acquisition costs for the public sector, without factoring in savings in procedural costs, since they are not available for research.
- A reduction in the number of intermediaries, as the system makes it possible for the producer to sell directly to the consumer (in this case, the Federal Government); an almost certain increase in the number of suppliers as technological innovation removes geographical limitations by making the process available to the entire country via the Web. Hence, as argued by Turban *et al.* (2000), in Web-based reverse auctions, the admission barriers are far less than for traditional auctions, thus leading to lower acquisition prices for the auctioneer.
- As a ripple effect, product prices tend to fall the greater the number of participants involved, and also, due to the transparency of the process, making it possible for suppliers to analyze whether they can bid a lower amount than the one listed on the Web, by lowering their bottom-line.

Efficiency involves achieving more for less; therefore this acquisition model is clearly efficient. However, only after an evolution in scale would it be possible to make a comparison between this procedure and others to be established in the future.

As regards efficacy, it is also possible to see that the on-screen procedure meets the desired objective, as it has brought about a drastic reduction of time involved, which is highly relevant as we are dealing with acquisition of pharmaceutical products. Usually, a public tender of this kind would last several weeks. In this case, only five hours and forty minutes were needed to accomplish the targets set by the MPAS.

Where accountability is concerned, the transparency of the procedure adopted is indisputable, as it guarantees access to information and real-time follow-up to all citizens, making it easier for society to control the application and use of public resources. All data is available on the Web and anyone may monitor any auction staged by the Government.

Regarding effectiveness (Checkland, 1981), the authors understand that the right thing has been implemented, though this implies a value judgment, so we will not dwell on this aspect.

Another highly significant result was the actual cost reduction of nearly 30% for acquisition of pharmaceutical products, in addition to the marked correlation between the number of bids placed and the cost reduction percentage, with a price reduction of nearly 1% for each bid entered.

6.2 Obstacles, Causes and Solutions

As previously stated, an explanatory approach was applied in the case study. Explanatory case studies are useful to assess how an intervention is working and why. The methodology verifies whether problems have arisen and modifications are needed in the intervention, and tries to explain the causal effects found. Different sites are needed in order to develop a comparative analysis (Morra and Friedlander, 1999). Subsequently, from direct observation and interviews with the auction participants and using a framework proposed by Joia (1998), one can create Table 11.1 that addresses the obstacles encountered during the whole process, the perceived causes for these obstacles and some possible solutions as suggested by the authors.

According to Markus (1983, p. 431), there are three theories to explain user resistance to information systems. The first theory addresses the fact that resistance exists because of factors inherent to persons or groups (Markus, 1983, p. 431). In this case, people may have some special characteristics that impede them from using the systems. According to this first theory, lack of training, resistance to technology, fear of computers, no perceived utility in the system, among others, can be considered obstacles preventing users from taking advantage of a new information system.

The second theory addresses the system itself. It may have flaws in its design. According to this theory, lack of flexibility, poor user-friendliness and unnecessary complexity, among others, can be considered sources of user resistance to adequate use of the system.

Besides these two vectors, people and systems can interact creating the third resistance theory to management information system implementation, according to Markus (1983) and Kling (1980). The key word in this case is interaction. It should be noted that this explanation identifies neither the system nor the organizational setting as the cause for resistance/acceptance, rather the interaction between the two.

Table 11.1
Obstacles, Causes and Solutions

Obstacles	Causes	Solutions
Structural		
Focus only on direct man-power and indices	Obsolete decision criteria	In-depth analysis of the costs and benefits involved
Failure to perceive the actual benefits	Lack of measurement of intangible benefits	Intangtible and tangible productive analysis
High risk for the managers	Reward system not considering innovation	Different reward systems for managers
Lack of co-ordination and co-operation	Organizational fragmentation	Systems to allow co-ordination/ co-operation
High expectation and hidden costs	Selling of an unreal system	Planning strategic objectives
Human		
Risk avoidance	Fear of change and uncertainly	Communication and in-volvement
Resistance	Fear of loss of power and status	Board engaged in project implementation
Unplanned decisions and fear of being made redundant	Orientation and action: lack of patience with planning	Pilot Project planning: long-range objectives
Technical		
Incompatibility of systems	Purchase of different hardware and software plaforms	Purchase of only one integrated system; write own system; neutral transfer files

Several distinct variations of the interaction theory can be identified (Markus, 1983, p. 431). One, called the socio-technical variant, focuses on the distribution of responsibility for organizational tasks across various roles and on work-related communication and co-ordination on the division of labour. New information systems may ascribe a division of roles and responsibilities, which are at variance with existing ones. They may structure patterns of interaction that are at odds with the prevailing organizational culture. In this light, systems can be viewed as a vehicle for promoting organizational change. Similar articulations of this variant of the interaction theory can be found in Keen (1980) and Ginzberg (1975).

A second variant of the interaction theory can be called the political version. In this case, resistance/acceptance is explained as a product of the interaction of system design features with the intra-organizational distribution of power and status, defined either objectively, in terms of horizontal or vertical dimensions, or subjectively, in terms of symbolism (Markus, 1983).

As can be seen in Table 11.1, the perceived obstacles/causes associated with the Human Dimension of this enterprise were mostly derived from the aforementioned first theory (the User theory). On the other hand, the perceived obstacles/causes associated with the Structural Dimension of this enterprise were derived mainly from the Interaction Theory outlined above, addressing both its Socio-Technical and Political versions. Finally, the perceived obstacles/causes associated with the Technical Dimension of this enterprise were linked to the system's characteristics.

Interestingly, the major obstacles/causes did not arise from the technology implemented per se, but rather from organizational and human issues. Based on that conclusion, it was possible to propose some solutions to be implemented. These solutions drew heavily on a deeper analysis developed by Joia (1998) in a similar setting, encompassing not only the technology itself, but also organizational and structural obstacles (see Table 11.1).

Thus, for technological implementation such as that presented in this research, a great deal of attention is required when dealing with new services to be enabled by the technology implemented, as well as with the processes that will be redesigned on account of these newly available services. Hence, according to Joia (1999), the following taxonomy must be closely monitored—technology⇒services⇒processes.

7. Closing Remarks

Despite being presented as a single case study, 128 observations were recorded in this research addressing auctions of different goods, so it can be said that an embedded single case study was undertaken, as multiple units of analysis were taken into account and measured. Construct validity was ratified by using several data sources; internal validity was achieved through the statistical analysis of the case; external validity was ensured as several auctions were analyzed, proving that these results can be replicated, *ceteris-paribus,* in other situations; and finally, the reliability of the case was confirmed by using a case study protocol, developing a data base and asking other people to review the paper.

From the legal standpoint, the *Pregão* is an improvement on the rules of participation for the Federal Public Administration. This new model opens the market up to increased competition and broader participation of players in public tenders, thus contributing to the overriding quest to cut expenses in line with the goal of fiscal adjustment. The *Pregão* guarantees immediate savings in the acquisition of goods and services, especially those involved with the costs of the federal administrative machine. This model also permits agility in acquisitions as bureaucracy is removed from legal procedures.

Finally, as stated by Lenk and Traunmüller (2001, p. 72): "There are good reasons to believe that the public sector cannot stay aloof as commercial business is undergoing deep changes. The technology potential itself changes widely-held fundamental concepts of good practice". This research constitutes a clear illustration of the above statement.

References

Ballantine J. A. and Cunningham N. (1999) Analyzing Performance Information Needs, in: Heeks, R. (Ed.) *Reinventing Government in the Information Age*, Routledge: London, 331–349.

Blackmor D. A. (2000) Where the Money is, *The Wall Street Journal*; R30, R32, April 17.

Braga, E. (2001) Governo vai às compras na rede, Rio de Janeiro. *Tema*, **153**, 30–34.

Brignall, T. J., Fitzgerald, L., Johnston, R., Silvestro, R. and Voss, C. (1992) Linking Performance Measures and Competitive Strategies in Service Businesses: Three Case Studies, in: Drury, C. (Ed.) *Management Accounting Handbook*, Oxford: Butterworth-Heinemann in conjunction with the Chartered Institute of Management Accountants.

Campos, A. M. (1990) Accountability: Quando poderemos traduzi-la para o português?, *Revista de Administração Pública*, **24**, 2, 30–50.

Checkland, P. B. (1981) *Systems Thinking, Systems Practice*. Chichester: John Wiley.

Checkland, P. B., Forbes, P. and Martin, S. (1990) Techniques in Soft Systems Practice. Part 3: Monitoring and Control in Conceptual Models and in Evaluation Studies, *Journal of Applied Systems*, **17**, 29–37.

Einsenhardt, K. M. (1989) Building Theories from Case Study Research, *Academy of Management Review*, **14**, 4, 532–550.

Fitzgerald, L., Johnston, R., Brignall, T. J., Silvestro, R. and Voss, C. (1991) *Performance Measurement in Service Businesses*. London: Chartered Institute of Management Accountants.

Fountain, J. (2001) *Building the Virtual State: Information Technology and Institutional Change, Washington*. DC: Brookings Institution Press.

Ginzberg M. J. (1975) Implementation as a Process of Change: A Framework and Empirical Study, Rept. CISR—13, *Center for Information Systems Research*, Cambridge: MIT Press.

Joia, L. A. (1994) *Reengenharia e Tecnologia da Informação: O Paradigma do Camaleão*. São Paulo, Brasil: Editora Pioneira.

Joia L. A. (1998) Large-Scale Reengineering on Project Documentation at Engineering Consultancy Companies, *International Journal of Information Management*, 18, 3, 215–224.

Joia, L. A. (1999) An IT-Based Taxonomy Model for Enterprise Engineering, *Information Strategy—The Executive's Journal*, **15**, 3, 27–35.

Kalakota R., Robinson M. and Tapscott, D. (1999) *E-Business: Roadmap for Success*, Addison-Wesley.

Kambil A. and Heck E. (1996) Re-engineering the Dutch Flower Auctions: A Framework for Analyzing Exchange Organizations, New York University, Department of Information Systems, *Working Papers Series, Stern IS-96–24*.

Keen P. (1980) Information Systems and Organizational Change, Rept. CISR—46, *Center for Information Systems Research*, Cambridge: MIT Press.

Kubicek H and Hagen M. (2001) Integrating E-Commerce and E-Government: The Case of Bremen Online Services, in: Prins, J. E. J. (Ed.) *Designing E-Government*, The Hague, The Netherlands: Kluwer Law International, 177–196.

Lenk K. and Traunmüller R. (2001) Broadening the Concept of Electronic Commerce, Prins, J. E. J. (Ed.) *Designing E-Government*, The Hague, The Netherlands: Kluwer Law International, 63–74.

Leitner C. (2003) e-Government in Europe: The State of Affairs, *e-Government 2003 Conference*, Como, Italy, 7-8 July, 33.

Machado U. (2001) O Estatuto das Licitações e Sua Evolução no Tempo, Rio de Janeiro. *Tema,* **155**, 26–28.

Maira A. N. and Taylor M. R. (1999) The Big Picture: An Overview of Electronic Commerce, www.adlittle.com/publications/prism-1q99.asp

Markus L. M. (1983) Power, Politics and MIS Implementation, *Communications of the ACM*, **26**, 3, 430–444.

McHale, J. (1972) *World Facts and Trends*, New York: Collier Books.

Mokate K. (1999) *Eficácia, Eficiência, Equidad y Sostenibilidad*, Instituto Internacional para el Desarrollo Social INDES, Junio.

Morra L. and Friedlander A. C. (1999) *Case Study Evaluations*, OED (Operations Evaluation Department) Working Paper Series No. 2, May, World Bank.

Neef D. (2001) *E-Procurement: From Strategy to Implementation*, Prentice-Hall PTR/Sun Microsystems Press.

OECD (1998) Information Technology as an Instrument of Public Management Reform: A Study of Five Countries, Public Management Committee: www.olis.oecd.org/olis/1998doc.nsf/LinkTo/PUMA(98)14

Osório, M. (2001) O que é o Governo Eletrônico, *Tema*, **153**, 30-42, Rio de Janeiro.

Prins, J. E. J. (2001) Electronic Government: Variations of a Concept, in: Prins, J. E. J. (Ed.) *Designing E-Government,* The Hague, The Netherlands: Kluwer Law International, 1–5.

Roebeke, L. (1990) Measuring in Organizations, *Journal of Applied Systems Analysis,* **17**, 115–122.

The Economist (2000) Government and Internet Survey, London, 24-30, June 22nd.

Turban E., Lee J., King D. and Chung H. M. (2000) *Electronic Commerce: A Managerial Perspective,* New Jersey: Prentice-Hall Inc.

Wurman P. R., Wellman M. P. and Walsh W. E. (2001) A Parametrization of the Auction Design Space, *Games and Economic Behavior*, **35**, 304–338.

Yin R. (1994) *Case Study Research: Design and Methods*, Thousand Oaks, California: Sage Publications.

About the Authors

Luiz Antonio Joia is an Associate Professor and MBA Co-ordinator at the Brazilian School of Public and Business Administration of Getulio Vargas Foundation. He is also an Adjunct Professor at Rio de Janeiro State University. He has published two books, several chapters and articles in journals such as: *Internet Research— Electronic Networking, Applications and Policy; International Journal of Information Management; Journal of Global Information Management; Journal of Intellectual Capital; Information Strategy—The Executive's Journal; Journal of Teacher Training and Technology; Journal of Knowledge Management; Journal of Workplace Learning.* He is a member of the Editorial Board of the *Journal of Intellectual Capital* (Emerald) and of the *Electronic Government: An International Journal* (Inderscience).

He holds a B.Sc. in Civil Engineering from the Militar Institute of Engineering, Brazil, and a M.Sc. in Civil Engineering and a D.Sc. in Engineering Management from the Federal University of Rio de Janeiro. He also holds a M.Sc. in Management Studies from the Oxford University, U.K. He was a World Bank consultant in Educational Technology and is an invited member of the Technical Board of the Working Group WG 8.5 (Informatics in the Public Administration) of the IFIP (International Federation for Information Processing).

Fuad Zamot is a professor of FGV Management in the area of Strategic Planning. He is also a member of the FGV study group—Court of Accounts for the Account School Project. He has a vast experience in teaching, having been a Professor of Strategic Planning at the Serzedello Corrêa Institute of Accounts Court of the State of Rio de Janeiro. He was an Associate Researcher of the Americas Study Center—CEAs at Candido Mendes University, and for a period he was the Project Manager in the municipal area with emphasis on the co-ordination of several organs of the public sector.

In reference to his Academic Background, Fuad Zamot holds a Master's Degree in Business Management from FGV, Post-graduate studies in Public Administration from FGV as well as a Degree in Law .

Fuad Zamot has had his work on Internet-Based Reverse Auctions in the Brazilian Government as well as on Informations Technology and Accountability in Accounts Court published in *Electronic Journals on Information Systems in Developing Countries* and in International Conferences on Electronic Commerce.

12. Decision Support Systems for Small Scale Agroindustrial Investment Promotion in Rural Areas[30]

Carlos Arthur B. da Silva

Agribusiness Management Program, Federal University of Viçosa, DTA-UFV
36571-000, Viçosa, MG, Brazil
carthur@ufv.br

Aline R. Fernandes

Brazilian Ministry of Science and Technology (MCT)
Brasília, DF, Brazil
afernandes@mct.gov.br

Abstract

A series of decision support systems (DSSs), developed as part of a government program to promote the implementation of small scale agroindustrial projects in rural areas, is described. The DSSs constitute an alternative to the use of printed agroindustrial profiles to disseminate information on technical and economic aspects of such projects. Users have access to information on the potential projects and can evaluate their feasibility considering their own data and relevant assumptions. Information is presented in hypertext and graphical formats, and in conventional data tables. The hypertext files provide descriptions concerning the technological processes, related legislation and potential markets, while graphical images present plant floor plans and 3-D views of facilities and equipment lay-out. Data is available on several items, including technical coefficients, equipment costs, raw material and other input costs, interest rates, other loan conditions, and product mixes and prices. They can be easily changed by the user, so as to better reflect specific conditions. Economic-engineering models allow the computation of standard financial evaluation indicators. Sensitivity analysis is also performed. Results can be either displayed on screen, saved or printed. 15 systems have been developed, covering a wide array of agroindustrial projects. Evaluation tests were performed with a sample of users. Results were extremely favourable, confirming the initial hypothesis of the potential of the proposed approach as an effective means to promote agroindustrial investments.

---

[30] An earlier version of this work has been presented at the 2nd Asian Conference for Information Technology in Agriculture, Suwon, South Korea, June 2000.

1. Introduction

Agroindustrial projects are frequently promoted by governments and development agencies as strategic components of rural development programs. In fact, as pointed out by Austin (1992), a considerable share of international aid flows in the 80's and early 90's was directed to investments in agroprocessing projects. Informal consultations with development experts suggest that these trends are even more prevalent today.

The reasons for such an interest in agroindustries as major forces in rural development efforts are associated with the potential benefits generated by investments in the sector. It is known that the multiplier effects of agroindustrial investments are amongst the highest, when compared with other economic sectors (Najberg and Pereira, 2004). By aggregating value to agricultural products, agroindustrial processing contributes to the creation of employment opportunities both before and after the farm gate, thereby reducing the serious problem of rural migration, which is so common in less developed economies. Moreover, agroprocessing technologies contribute to the improvement of quality of raw materials and final products. Better quality, in turn, promotes improved health and nutrition and yields access to more demanding markets, both locally and internationally. Another important benefit is the availability of simple, small scale technologies, making a wide array of processing alternatives accessible to small investors.

In 1998, the Brazilian Ministry of Agriculture established a major effort to promote agroindustrial investments by groups of small farmers—the "Pronaf–Agroindústria" Program (Ministério da Agricultura e do Abastecimento, 1998). Its main objective was "...to contribute to the improvement of living conditions in rural areas, by fostering and supporting the integration of family farm associations to the agribusiness economy". The program offered credit lines for investments in new agroindustrial plants. Training and technological support and information systems to support planting, processing and marketing decisions were also provided. This program was later transferred to the Ministry of Agrarian Development and currently has a planned budget of US$250 million. The aim is to promote the creation or upgrade of some 7,000 small scale agroindustries over the 2004–06 period.

In order to present potential investors with basic information on technologies, as well as on costs and benefits of alternative enterprises, the so called "agroindustrial profiles" are usually prepared by governments and development agencies. These publications briefly describe the enterprise, including information on the processing technology, the needed equipment and buildings, labour requirements, potential markets and cost-benefit issues. As part of its support actions, the "Pronaf–Agroindústria" Program originally planned the development of a series of such profiles.

However, one critical difficulty with the "profile" approach is the static characteristic of the information conveyed. In fact, for a country with marked regional differences as Brazil, it is not reasonable to assume that input costs, product prices, product mixes and other key agroindustrial project hypotheses would hold true nationwide. Furthermore, with the historic inflation problems of the country, there would be a non negligible risk that relative and absolute prices could change, thus drastically reducing the usefulness of the provided financial information.

To circumvent these perceived shortcomings, a computer based alternative to the traditional "profile" has been suggested by the authors—the "interactive agroindustrial profile" (Silva *et al.*, 1998). The interactive profile concept entails the development of decision support systems (DSSs), containing all descriptive information of the agroindustrial projects, as well as economic-engineering models that allow users to adjust project assumptions so as to reflect their specific conditions.

In co-operation with the Brazilian Ministries of Agriculture and Agrarian Development, 15 interactive profiles were developed for the aforementioned Agroindustrial Promotion Program, covering the following small scale agroindustrial projects—cheese production, cashew-nut processing, poultry slaughter, hog slaughter and processing, brown sugar, sugarcane brandy distillery, fruit pulp, soybean meal and oil, milk cooling centres, diversified milk processing, goat milk cheese, manioc flour (two plant sizes), banana drying and minimally processed vegetables. A specific system was developed for each of the alternative projects. The general structure of these DSSs, which were named "SAAFI–Agro" (Portuguese acronym for "Decision Support System for Financial Evaluation of Agroindustrial Projects") is discussed next.

2. System Description

Two related objectives were sought. One was the provision of a user friendly environment for the dissemination of agroindustrial technologies to potential investors. As such, descriptive and visual information were to be made available, warranting straightforward consultation. A second objective was the provision of interactive cost-benefit information, in the form of direct access to tables containing all basic data on project investments, costs, revenues and financing conditions, so that the provided figures could be readily changed by users. Once the desired changes are performed, the system should automatically compute investments, costs, revenues and cash flows. These figures in turn should be automatically considered in the computation of financial indicators, such as the internal rate of return, net present value, payback period and break even point. Finally, a sensitivity analysis, and an evaluation of the effects of under capacity utilization (learning curve analysis) should be conducted.

The systems were developed for the Windows 95/98 environment, with the Delphi 3 programming language. A traditional prototyping methodology (Turban and Aronson, 1998), was followed in the DSS development process.

The same basic structure was followed for all of the developed DSSs. An opening screen (Figure 12.1), offers access to a pull-down menu, by means of which consultation to a standard project can be made. Optionally, users can create new projects, by modifying the assumptions and data of the standard project, or retrieve previous modifications, saved from prior utilization of the system. Once a consultation option is made, users can access the descriptive and visual information, the cost-benefit information or the financial evaluation results.

Figure 12.1
Opening Screen of the DSS for the Banana Drying Agroindustrial Plant

2.1 Descriptive and Visual Information

Descriptive information is mainly composed by a hypertext file presenting a characterization of the technical aspects of the enterprise. Information is offered on the processing steps, the raw materials needed, the necessary equipment, labour needs, building specifications, legislation, waste treatment processes and equipments, and quality assurance systems, among others. In order to facilitate the consultation, most technical terms are linked to explanation boxes. Complementing the technical texts, processing flow charts, plant floor plans and 3-D plant views (Figure 12.2) are visually displayed. Print-outs of all texts and drawings can be easily obtained.

Figure 12.2
3D View of the Agroindustrial Plant (Banana Drying)

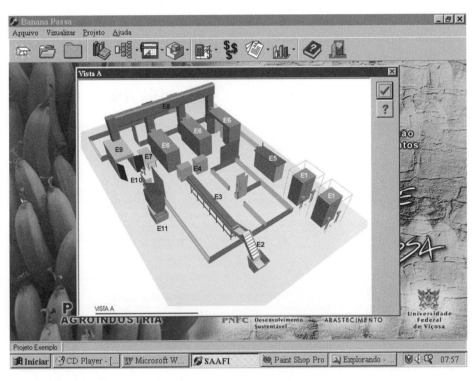

It should be noted that all descriptive and visual information provided correspond to a given plant size. This information set can not be altered by users, for it reflects technological standards and scale of operations considered appropriate for the purposes and target beneficiaries of the "Pronaf–Agroindústria" Program. On the other hand, changes can be performed in the cost-benefit modules.

2.2 Cost-Benefit Information

In accordance with the typical investment analysis methodologies, before a project's cash flow can be calculated, a cumbersome process for estimating investments, costs and revenues must be performed. Data must be gathered on the items that compose each of these categories, and the appropriate tables must be built. In addition, several aggregation steps must be taken, before the needed information for financial evaluation can obtained. The developed DSSs here described greatly simplify this process, by providing access to all relevant tables and by automating the aggregation steps and other necessary computations.

The cost-benefit components are organized in four main groups of data tables— fixed investment and working capital estimation, cost estimation, revenue estimation and financing alternatives. The first group contains figures on costs of equipment and other fixed investment items, such as land, vehicles, buildings (civil works) and others. The second one lists all input and raw materials needed, as well as other fixed and variable cost items such as labour, energy, depreciation, taxes, etc. Revenue estimation is performed by establishing a product mix and a set of selling prices. The financing schema combines information on interest rates, grace period, repayment schedules and others. Figure 12.3 illustrates the presentation format utilized.

Figure 12.3
Presentation Format for the Financial Data Tables

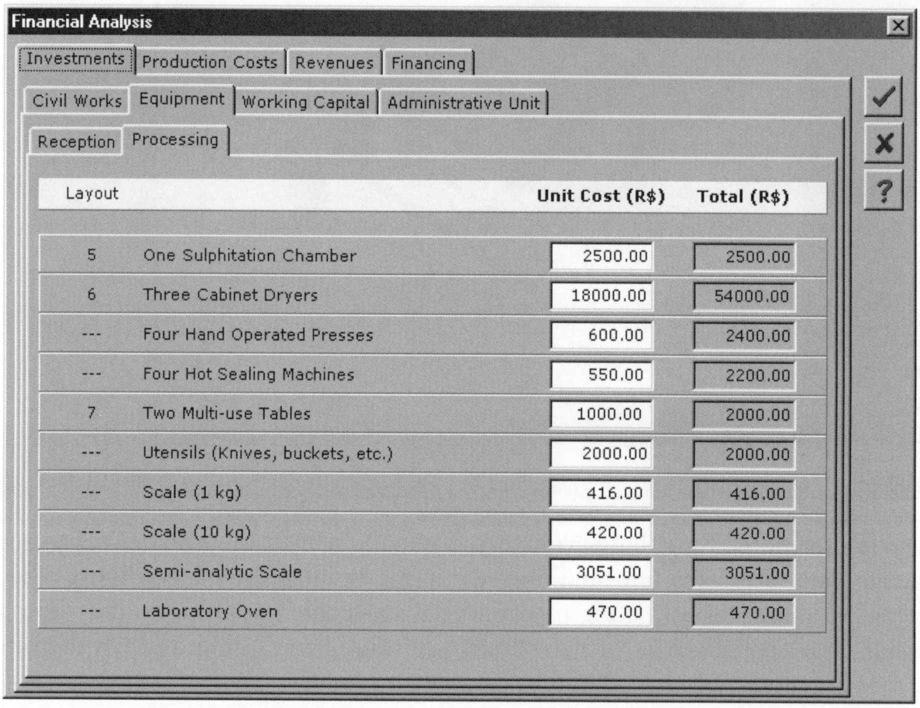

The screen displayed shows a list of equipment items and the respective investment estimates. The figures presented in the boxes with a white colour background can be readily changed by users, so as to adapt the project to their specific areas or conditions. Alternatively, these changes can be made as a way to assess the impact of key financial assumptions on the project profitability (what-if analysis). After the intended changes are made, it is then possible to access the financial analysis information component.

2.3 Financial Analysis

Besides presenting the project's cash flow and the standard financial evaluation indicators—net present value, internal rate of return, pay-back period and break-even point—the financial analysis component presents graphs with cost breakdowns, break-even charts, sensitivity analysis spiderplots (Figure 12.4) and "learning curve" charts.

Figure 12.4
Sensitivity Analysis Spiderplot

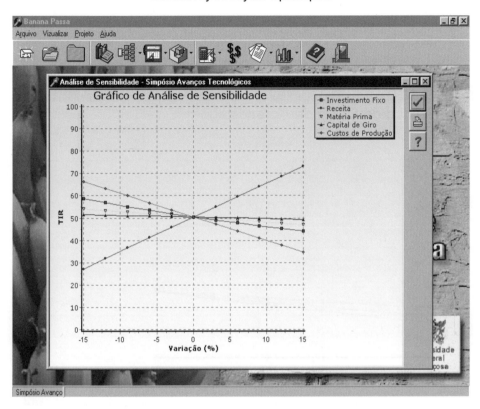

The spiderplots (Eschenbach, 1992) show the effects of altering key project assumptions on variables such as investments, costs and revenues. The learning curve chart (Figure 12.5) depicts the effects of running the agroindustrial plant below full capacity during one, two or three years of operation.

Concluding the utilization of the system, a complete print-out can be provided, in a format that contains all the necessary information for presentation of the project to a financing agency.

3. System Utilization

The developed DSSs have been widely distributed by the Brazilian Ministry of Agrarian Development, both by mailing CD-ROM's to potential users and by allowing free-of-charge downloads from a Ministry web page. Primary targets for the distribution efforts have been the agricultural extension services of Brazil's 27 states, cooperatives and other small farmers associations, departments of agriculture at the state and county levels, non-governmental organizations, banks and other financing institutions, as well as research and academic institutions.

Figure 12.5
"Learning Curve" Chart

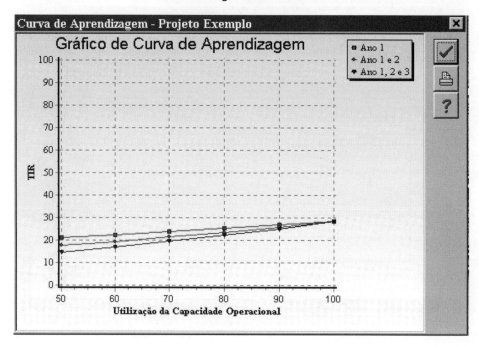

A survey with a sample of users is currently being performed. Preliminary results with 51 respondents of a mail questionnaire are confirming previous informal findings about the effectiveness of the developed systems. A substantial proportion of these users (71%) have declared their complete satisfaction with the software. Among the respondents, 43% have effectively utilized the systems to prepare agroindustrial projects, while 36% have used them as reference materials to provide technical assistance. Other alternative uses have been the utilization of the equipment specifications and plant lay-outs for the preparation of projects (21%), the utilization of the financial tables to help in the evaluation of alternative projects (22%) and the utilization of the agroindustrial process described as guides to implement processing techniques (21%).

Users were asked to present suggestions to improve the systems, but only 16% declared that enhancements were needed. In this regard, the main perception is the need to make plant sizes flexible, thereby allowing different scales to be evaluated for any given agroindustrial project. However, given the requirement that the technical and financial aspects of a project be associated with its planned scale of operations, this suggestion can not be easily implemented. An alternative is to provide, for each project file, a fixed number of plant size alternatives. In fact, we followed this approach in at least one of the DSSs, the cashew nut processing project, which considers 4 alternative plant sizes. Further developments of the systems will likely consider more than one plant size option in each DSS.

Finally, there were suggestions to prepare new systems for a sizable number of processing technologies of interest to small farmers, including goat slaughterhouses, honey processing, grape processing, corn flour and other corn based products, fish processing, coffee processing, jams and other fruit preserves, as well as sausages and other meat based products.

4. Conclusions

The preliminary survey results and the anecdotal evidence on the applications of the software we have gathered so far indicate that the proposed concept is workable and can be a very effective alternative to the traditional agroindustrial profile format. Indeed, we have already been notified of a number of small scale agroindustries that have been implemented based on the projects presented in the DSSs. Moreover, the Ministry of Agrarian Development is planning to expand the titles of the software series, largely by following the suggestions presented in our user survey.

Another indication of the interest in the proposed approach was the proposal presented by the Food and Agriculture Organization of the United Nations (FAO) for a partnership with the authors, towards the development of further decision support tools for agroindustrial project preparation and evaluation. The result of this partnership was the development of *Agriventure*, a flexible decision support tool that can be used to prepare and evaluate any type of agroindustrial project (Silva *et al.*, 2003). *Agriventure* can be freely downloaded from the FAO's web site, in both English and Spanish versions (www.fao.org/inpho).

In sum, the evidence we have suggests that the decision support systems developed are effective tools to promote agroindustrial investments and to disseminate technological and financial information on alternatives to aggregate value to agricultural production. Interested readers can download the software (in Portuguese) from the Ministry of Agrarian Development's web site (www.pronaf.gov.br) or from the Federal University of Viçosa's web page (www.ufv.br), under the link labelled "softwares disponíveis" (available software).

References

Austin, J. (1992) *Agroindustrial Project Analysis*. Washington D.C. The World Bank

Eschenbach, T. G. (1992) Spiderplots vs. Tornado Diagrams for Sensitivity Analysis, *Interfaces*, **22**, 6, 40–46.

Ministério da Agricultura e do Abastecimento—MA (1998) *PRONAF-Agroindústria: Integração, Agroindustrialização e Comercialização da Produção da Agricultura Familiar*. Brasília, DF

Najberg, S. and Pereira, R. De O. (2004) Novas Estimativas do Modelo de Geração de Empregos do BNDES, *Sinopse Econômica* **133.** www.bndes.gov.br/conhecimento/estudos/estimativas.pdf

Silva, C. A. B., Fernandes, A. and Jallad, J. (1998) *Perfis Interativos: Uma Proposta para a Disseminação de Tecnologia e Fomento à Implantação de Empreendimentos Agroindustriais*, ANAIS do Simpósio sobre Avanços Tecnológicos na Agroindústria Tropical (Separata). Fortaleza, CE

Silva, C. A. B., Perez, R., Fernandes, A. R. and Machado, J. (2003) Agriventure: A Decision Support System for Project Preparation and Evaluation, *Proceedings of the IV Congress of the European Federation for Information Technology in Agriculture*, Debrecen, Hungary, July, 67–72.

Turban, E. and Aronson, J. (1998) *Decision Support and Intelligent Systems: Management Support Systems*. Prentice Hall, Upper Saddle River, New Jersey.

About the Authors

Carlos Arthur Barbosa da Silva is a Professor at the interdisciplinary Agribusiness Management Program of the Federal University of Viçosa, in Brazil. Holding a Ph.D. degree in Agricultural Economics from Michigan State University (USA), he teaches undergraduate students in the areas of Agroindustrial Projects and Dairy Plant Management. He teaches also Decision Support Systems and Research Methodology at the graduate level. His research interests are focused on modelling and decision support applications in managerial problems of concern to food and agribusiness firms. He is a frequent consultant to the Brazilian Ministries of Agriculture and Agrarian Development, non governmental organizations such as the Brazilian Small Business Promotion Agency, private companies and international organizations, such as the Food and Agriculture Organization of the United Nations. Carlos was the first President of the Brazilian Society of Information Technology in Agriculture and Agroindustry, where he started the Brazilian Journal of Agroinformatics. He has co-authored two books on Agroindustrial Projects and one on Microcomputer Applications in Development Environments. He has also published extensively in the *Brazilian Journals of Agricultural Economics*, *Food Technology and Dairy Technology*, among others. His works appear in the proceedings of most international conferences organized by the European and Asian Federations of Information Technology in Agriculture over the last 10 years.

Aline Regina Fernandes recently received her doctoral degree in Food Science and Technology at the Federal University of Viçosa, in Brazil. She has a Food Engineering B.Sc. degree and her graduate studies had a multi-disciplinary focus, covering the areas of food processing, agribusiness economics and management and decision support systems, with studies developed also at Michigan State University and the Massachusetts Institute of Technology (MIT) in the USA. Her dissertation research consisted of a systems simulation analysis of the operational dynamics and economic sustainability of small scale agroindustrial processing firms. She has an extensive professional experience in the formulation of agroindustrial projects, mostly for small scale enterprises, and has co-authored two books and dozens of journal articles and technical reports in this area of activities. Aline has participated in the development of the series of decision support systems that are being distributed by the Brazilian Ministry of Agrarian Development in order to promote investments in small scale agroindustrial plants. She currently works as a Science and Technology Analyst for the Brazilian Ministry of Science and Technology, where she is involved in policy formulation and project evaluation, mostly for the food and agribusiness area.

Information Technology Infrastructures for Developing Countries— Key Concepts and Cases

Sajda Qureshi

It is accepted that IT infrastructure has brought about new ways in which developing countries are able to harness their own resources to provide services to their constituencies and foreign companies. Some of us even suggest that IT may be a driver for economic development and may bring about measurable improvements in the lives of people (Qureshi, 2004). However, it is still not clear what are these changes brought about by IT infrastructure? Moreover, for better or for worse, what effects does this IT infrastructure have on the lives of people? The papers in this section attempt to address parts of these questions and offer insights into factors that effect the success of countries attempting to harness the potential of their IT infrastructures.

The first paper in this section is entitled: "Realizing the Development Potential of North-South Business Process Outsourcing—The Case of Fiji" by Charles Davis, Jim McMaster and Jan Nowak. It examines the concepts of IT Enabled Business Services (ITEBS) and business process outsourcing. The authors suggest that Information and Communication Technologies have the effect of unbundling the production from the consumption of services. This means that services can be traded regionally and internationally as they are produced in one region and delivered in another. Outsourcing is a very well known example of this unbundling of services as a result of ICT infrastructure. In particular, the fastest growing area of outsourcing identified by the authors is business process outsourcing. It is in this area that the authors suggest that Fiji may have potential in the development of IT enabled service industries. They propose low wages, English language capabilities, education levels and requisite IT skills to be key conditions effecting success in ITEBS. In addition to low wages and an English speaking work force, Fiji has low inflation, a relatively stable foreign exchange rate and a reliable set of enforced contract laws that make it an attractive choice for ITEBS. However, there is a lack of adequate IT skills, the cost of internet access is high as there is only one internet service provider in Fiji and the bureaucracy facing foreign investors is high. In the face of recent back office developments, the authors provide policy recommendations to enable small island states such as Fiji can attain competitive advantage through ICTs.

The second paper entitled: "The Provision of Internet Services in India" by Peter Wolcott considers the market for Internet services in India. Following a description of the evolution of IT policy and the growth of the Internet, this paper reveals the tension between policy makers, the incumbent state-owned telecommunications providers and the new Internet Service Providers (ISP) that have fuelled the growth of internet infrastructure in India. Using a framework for analyzing strategies for ISPs, the authors suggest that the ISPs in India may be categorized into the incumbent telecommunications companies and the small and regional ISPs. They suggest that there is a competitive balance to be gained as the incumbent telecommunication companies may enjoy competitive advantages through supplying of POPs, established relationships with suppliers, cost advantages and economies of scale. Small ISPs may enjoy competitive advantages through the provision of customized support services. However, incumbent ISPs use relatively old technology and may face threats from new entrants in the market for backbone capacity and the regional ISPs are compromised in the dependency on the incumbent telecommunications for infrastructure. The author concludes that while Internet service provision has made enormous strides in India, only 1.5% of the population use the Internet.

The third and final paper in this section is entitled" Information Technology for Development in India—The Kerala Experience" by K. G. K. Nair and P. N. Prasad. This paper considers the development of IT infrastructure in the government of the state of Kerala in India. The IT policy of the Kerala state government focused on decentralization of technology starting at the grassroots. A number of local computerization projects in local government are reported in this paper. The results of computerization have been promising, time for the delivery of services has been significantly reduced and processes simplified. According to the government of Kerala's "rights of way" policy, infrastructure providers laying fibre optic cables may use public properties for the construction of a fibre optic backbone across the state. This infrastructure is to ensure that high speed connectivity is available in all district headquarters. The high level of investment in IT infrastructure and IT skills has meant that software exports have risen in Kerala, although not as high as exports of its neighbouring states. The authors suggest that the state's poor performance in the IT industry can be attributed to its industrial backwardness. Despite investment in industrial parks and IT infrastructure, foreign and domestic investment is lower than expected leaving facilities underutilized. The authors suggest that strategies for promoting investment and utilization of the IT infrastructure should be carried out.

13. Realizing the Development Potential of North-South Business Process Outsourcing—
The Case of Fiji[31]

Charles H. Davis

Ryerson University
Toronto, Canada
c5davis@ryerson.ca

Jim McMaster

University of the South Pacific
Fiji Islands
mcmaster_j@usp.ac.fj

Jan Nowak

University of the South Pacific
Fiji Islands
nowak_j@usp.ac.fj

Abstract

Business process outsourcing (BPO)—outsourcing of information systems, data processing services, and other IT-enabled business services—represents a huge development opportunity for the South, and many developing countries or regions have targeted development of delivery capability of entry-level IT-enabled business services (ITEBS) as a strategic priority. IT-enabled business services are increasingly footloose and can be attracted on the basis of the cost advantages, infrastructure, or amenities provided by a location, depending on the segment in question. Geographic concentrations of such services can drive demand for improved IT infrastructure, technical training, and specialized services from local suppliers.

This chapter examines the options available to Fiji for participating in the international trade in IT-enabled business services and business process outsourcing.

---

[31] Authorship listed in alphabetical order. This is a revised version of Davis *et al.* (2002).

The chapter describes the segments of IT-enabled business services in terms of their skill level and the value they add, and discusses current trends in the international trade in business services, with particular emphasis on North-South ITEBS outsourcing. The chapter also identifies conditions that low-income countries must create in order to develop entry-level ITEBS export capability. In this context, Fiji's potential to develop or attract such services delivery capability is analyzed, including its labour cost, linguistic, geographic, educational, and other location advantages, as well as telecommunication infrastructure quality and cost disadvantages. The analysis covers a number of IT service industry projects recently initiated both by the Fiji Government and private sector enterprises. The chapter concludes with a critical review of the domestic regulatory environment affecting IT-enabled services development and with recommendations aimed at nurturing development of ITEBS export capability in Fiji.

1. Introduction

Global trade in IT-enabled services (ITES) is expanding rapidly as connectivity decreases the transaction and communication costs among firms. Chief among these services are IT-enabled business services (ITEBS)—services that are used internally by firms to produce a final good or service for customers. Lower-skill ITEBS, such as keyboarding, text entry, transcription, data processing and contact centres can be located successfully in low-income countries, provided that infrastructure standards and other conditions of service quality are met. India is the undisputed leader in business process outsourcing, but other countries with major outsourcing capability include Canada, China, the Czech Republic, Hungary, Ireland, Israel, Mexico, the Philippines, Poland, Russia, and South Africa. Belarus, Caribbean states, Egypt, Ukraine, Bangladesh, Cuba, Ghana, Senegal, and quite a few others, including Fiji, are developing outsourcing capability or have announced their intention to do so (Rundell, 2003; UNCTAD, 2003). Highly knowledge-intensive business services (such as R&D and engineering, software development, content production, or highly reliable applications hosting) have been traditionally located in or near major metropolitan areas in developed countries, but they are increasingly footloose and now can be found in locations with pools of highly skilled workers and appropriate infrastructure and amenities.

This chapter discusses the options available to a developing country, Fiji, to participate in the exploding international trade in IT-enabled business services. The authors first analyse global trends in tradable IT-enabled services, focusing especially on the migration of the entry-level portion of ITEBS from developed to developing countries. Then they identify the technological, human resource, political, and business-regulatory conditions that developing countries must create or be endowed with in order to be providers of outsourced ITEBS. Based on this identification of conditions, Fiji's potential for the development of ITEBS is analyzed and policy recommendations put forward.

The chapter is based on an extensive review of current literature and reports dealing with ITES and export development, and draws upon several individual country cases reported recently. As one of the authors has been involved in advising the Fiji Government on the service sector development, the chapter calls upon this author's intimate knowledge of the Fiji business environment and Fiji Government's strategies and policies.

2. Trends in Services—Disaggregation, IT-enablement and Trade

The service sector encompasses transportation, travel, communication, construction, insurance and financial services, computer and information services, royalties and licence fees, government services, personal and recreational services, and other business services. The rapid growth of international trade in services is a significant feature of contemporary economic development. United Nations Conference on Trade and Development (UNCTAD) estimates that world trade in services grew at an annual rate of 6.6% between 1990 and 2000. Developing countries' share of international trade in services grew at 10.1%, and their share of service exports increased from 15.7% to 21.2% between 1990 and 2000, representing income of more than $300B (UNCTAD, 2002). The fastest growth in internationally traded services occurred in computer and information services (*ibid.*).

The General Agreement on Trade in Services (GATS) recognizes four modes of service delivery: (i) supply across borders (as when services are delivered remotely); (ii) consumption abroad (as when customers travel from abroad to consume services delivered locally); (iii) commercial presence (as when the service provider establishes a physical presence in an export market); and (iv) travel by natural persons to an export market to deliver services (UNCTAD, 1998). Generally speaking, ITEBS (mode 1) are tending to substitute for other modes of service delivery, enabling the development of markets for these services to proceed faster than policy-enabled liberalization of service trade in other modes.

Driving the expansion of tradable services are advances in information and communication technologies (ICTs). Traditionally services were regarded as non-transferable and non-storable, requiring joint production between producer and consumer. In other words, services were essentially non-tradable. However, the diffusion of ICTs within the business sector lowers the costs of transacting, communicating, and co-ordinating among business units, making it possible to disaggregate or un-bundle the production and consumption of information-intensive service activities. This "blows to bits" value chains and induces the emergence of markets for services[32].

Specialized slivers of value production can be delivered remotely from places that provide suitable infrastructure and personnel at suitable cost (Quinn, 1992; Apte and Mason, 1995; Casson and Wadeson, 1998; Evans and Wurster, 2000; Wymbs, 2000). Un-bundling or disaggregation of information-intensive services can separate production from consumption and permit previously non-tradable services to be ac-

---

[32] This process is also referred to as "splintering" or "disembodiment" of services from goods production (Miozzo and Soete, 2001).

tively traded regionally and internationally. Services can be remotely produced and delivered when they do not require manipulation of physical objects or close interaction with the customer and when they are data-intensive (Apte and Mason, 1995; Miozzo and Soete, 2000).

Demand for outsourced services is soaring, spurred by opportunities to reduce the costs of production. The economic slowdown in industrialized countries has increased the incentives to outsource non-core activities in order to obtain greater efficiencies (Corbett, 2001b). Overall, global outsourcing of manufacturing and services doubled in value to approximately one trillion dollars between 1997 and 2000, with North America, Europe, and Asia accounting for 94% of the outsourcing market (Corbett, 2001a). The fastest growing areas of outsourcing are in business process or back-office functions such as human resource administration, media management, information technology, customer care, and marketing (Corbett, 1999). The value of business process outsourcing (BPO) is expected to be between $122B in 2003 and $240 in 2005 (CyberAtlas, 2003).

Outsourcing and procurement have become strategic issues for the firm, which must determine which steps of the value-adding process it will produce internally and which kinds of business relationships it must maintain with suppliers of essential and commodity inputs. Services may be insourced or outsourced from a single location, from multiple domestic locations, or from multiple global locations (Apte and Mason, 1995). Global multi-location insourcing and global outsourcing are the two development paths open to new entrants in the ITEBS business[33]. Many firms are establishing wholly-owned subsidiaries in lower-cost regions that take advantage of labour costs, financial incentives, and in some cases pools of highly specialized workers to provide skilled back-office services such as claims processing, R&D, or logistics management. For example, GE Capital India is the largest ITBES provider in India with 10,000 employees. It provides accounting, claims processing, credit evaluation, and other services to 80 GE branches around the world (Elsham, 2001). Global outsourcing occurs when firms make arm's length purchases of remotely-delivered services. Countries that want to enter the ITEBS business need to consider ways to attract service-providing subsidiaries through foreign direct investment as well as ways to develop or attract service exporters.

IT-enabled business services can be classified into three kinds of activities: administrative, customer services, and technical, and into three levels of skill- and knowledge-intensity—low, medium, and high (McMaster and McGregor, 1999). The resulting taxonomy, shown in the Appendix, provides a view of the ranges of service activities that can be offered at the three levels of complexity. The simplest tasks are routine data entry, customer service, and clerical activities. Intermediate services include ones requiring some judgement or unscripted interaction on the part of workers—

---

[33] Global multi-location in-sourcing is a form of intra-firm trade that is not reflected in ITES export statistics. Along with franchising it is the predominant form of service internationalization (Miozzo and Soete, 2001).

secretarial work, application or claim processing, management of records, transcription of specialized documents, and some kinds of website design and management. High-end ITEBS include remote delivery of professional services, dispute resolution, and complex technical or creative work such as software development, technical writing, animation, or remotely delivered educational or health services. Specialized markets are rapidly developing for low-skill services such as data processing and customer service delivery as well as for critical knowledge-intensive business service activities such as R&D and engineering design (Quinn, 2000). When business processes are of a strategic nature, firms prefer to outsource to "captive service farms" rather than to third-party service providers in an arm's-length relationship (Aron and Singh, 2002).

The use of information and communication technologies to make services tradable over long distances provides a new opening in an international economy that only a few years ago seemed to offer few development options to poor countries (Primo Braga, 1996). Export of services provides the principal opportunity for "development after industrialization" (Kobrin, 1999). Low-skilled keyboarding, text entry, data processing, transcription, translation, secretarial services, insurance claim processing, and customer interaction services are increasingly located in areas with lower labour costs and acceptable infrastructure in offshore information processing centres in Mexico, the Caribbean, Taiwan, the Philippines, India, and China. Lower-cost regions in economically developed areas such as Ireland and New Brunswick join countries such as India with pools of highly competent technical workers to compete for remotely-delivered knowledge-intensive business services such as engineering and design services, animation, data conversion, database development, accounting and auditing, distance education, network management, applications hosting, software development, and online health service delivery.

3. Factors Conditioning the Development of IT-enabled Service Industries in Low-income Countries

Low-end IT-enabled business services have been targeted as a strategic priority by many developing countries or regions. ITEBS are increasingly footloose, and any country with an appropriate telecommunications infrastructure and suitably qualified labour can compete for them. Entry barriers are relatively low—investment requirements are not great, the services are labour-intensive, cycle times are short, and many kinds of ITEBS do not require high levels of technical expertise. Therefore competition is intense. Here we briefly review the factors that condition the ability of a country to supply ITEBS exports.

Migration of business services to low-income countries is driven first and foremost by the lower costs of critical human resource inputs. Labour shortages in the North and labour costs in developing countries are the principal reasons that U.S. firms outsource their IT work (Carmel and Agarwal, 2000). Labour is often the single largest cost component of a service activity, representing up to 80%of the cost of a contact centre, for example. The cost savings for professional services supplied from a low-income country can be substantial. Qualified accountants in India are paid $3,000 per annum compared to $35,000 in the United States. Western companies such as GE Capital Services, British Airways and American Express are reported to have saved 40–50% of operational costs by shifting their customer interaction centres to India (Anonymous, 2001b). However, in addition to the cost savings that translate into shareholder value, executives of firms that outsource business processes or back office functions consider that outsourcing yields improved service quality and freedom to focus on core competencies (Management Trends in Outsourcing, 2001).

Linguistic ability is the second most important factor in the ability to compete as an ITEBS provider. Many customer services require knowledge of English. Countries that possess disciplined and literate workers able to work in or with English have a competitive advantage, at least in many of the lower-skilled service segments. This is one reason why certain Asian countries are best positioned to take advantage of the current outsourcing boom (Corbett, 1999).

Quality of telecommunications infrastructure and the connectivity speed and costs constitute the third most critical factor in developing successful ITEBS. As ITEBS are traded over long distance, they are very sensitive to the speed and costs of transacting and communicating. Access to high speed Internet connections and the costs of Internet services are of particular importance to ITEBS providers.

The fourth most important factor is the regulatory environment affecting the development of ITEBS. International rules of trade in services are sensitive to ongoing negotiations concerning movement of persons, definitions of subsidies, government procurement practices, taxes and regulations on electronic commerce, and market access. The elements of a national regulatory environment that affect the development of tradable ITEBS include "cyber laws" regarding digital signatures, information privacy, encryption, intellectual property; labour laws permitting contingent, twenty-four hour labour employment; regulations affecting the availability, cost, and quality of telecommunications services; taxation laws; and domestic, inward, and outward investment policies. Investors in ITEBS prefer to establish ventures in countries that offer:

- Transparent, consistent and predictable commercial laws and business environment,
- Sound macroeconomic management of the economy, with low inflation and relatively stable foreign exchange rates as well as easy repatriation of profits and capital,
- Safety and security of persons and property,
- Protection of property rights and enforcement of contracts, and
- Political and economic stability (Duncan *et al.*, 1999).

Private investment is also adversely affected by unpredictable changes in government policies that are perceived to have a negative effect on the profitability of investment (Serven, 1997; Burnetti *et al.*, 1997a, 1997b). Burnetti *et al.* (1997b) surveyed 3,600 enterprises in 69 countries to assess the critical factors that influenced private investment. Firms are deterred from investing in long-term business ventures if they cannot be sure which business regulations will apply in the future or how they will be interpreted by government officials. Also foreign investors need to be certain that the law courts will enforce contracts and protect their property rights. The investors' perception of the degree of political stability, predictability of judicial enforcement, and corruption are closely linked with investment (*ibid.*).

Even if firms enjoy a cost advantage to supply ITEBS internationally and operate in a favourable regulatory environment, they may not realize the potential for becoming competitive providers of such services. The key enablers of competitiveness are less the cost and regulatory environment than service quality, marketing capability, and credibility (Riddle, 2000). Firms must establish and maintain their credibility as service providers. They can do this by maintaining service quality and by using websites as proxies for quality. Firms must develop visibility and recognition, including possible national branding for service quality. They must learn to adopt culturally appropriate behaviour for customers, including meeting self-service and satisfaction expectations of customers, which may involve empowering staff in ways that are culturally unfamiliar (*ibid.*). They must have the administrative skills and infrastructure capability to bid on contracts, purchase, and receive and make payments in ways that customers prefer. National export promotion policies must provide services for exporters, including website promotion, business education, and electronic directories (*ibid.*). National SME-support policies must accelerate e-business enablement and ensure SME service exporters can have recourse to an effective local technical supplier industry.

Service quality has become a major issue in outsourcing and a likely major differentiator between successful and unsuccessful service exporters. Dun & Bradstreet estimates that between 20% and 25% of outsourcing relationships fail in any two-year period and that fully half of the relationships fail within five years (Dun & Bradstreet, 2000). The most important causes of outsourcing failure are the outsourcing supplier's lack of understanding of the customer's requirements, unexpectedly high costs, and poor service (*ibid.*). In such conditions, supplier accreditation is likely to assume increasing importance. Researchers at Carnegie-Mellon (Hyder *et al.*, 2001) have released an eServices Capability model that provides ITEBS outsourcing service providers a set of practices and standards designed to enable them to manage outsourcing relationships. The model covers the pre-contract, contract, and post-contract phases of the relationship and addresses organizational management, personnel, service design and delivery, technological infrastructure, and organizational knowledge issues (*ibid.*).

Although demand for ITEBS is booming, several factors on the horizon could reduce or alter the composition of demand for these services. Since demand for cost efficiencies drives the development of many outsourced ITEBS, technological advances that substitute for low-skilled labour may reduce the need for some kinds of services.

For example, smart products and optical recognition may reduce the need for data keyboarding, and voice recognition and artificial intelligence technologies may reduce the need for low-skilled customer service representatives. The differentiation of customer services along a scale from routine to high touch/high quality has led to the option of "near-sourcing" high grade customer contact services in Canada, where the labour force is literate, disciplined, and affordable (McCracken, 2003). Finally, the development of intelligent systems will allow firms to selectively route tasks on the basis of cost, opportunity, or skill, resulting in virtual service networks in which individual service providers can be located practically anywhere. In the North, an ample contingent workforce may be found for these jobs among younger people who have not been able to enter a career track in stable institutions, among older people with insufficient retirement income, or even among prisoners. Policymakers may find it appealing to offer incentives to locate virtual jobs in lagging regions within the national or regional economy. Furthermore, data security concerns may make it risky to locate ITEBS outside a national or regional perimeter. Thus future opportunities to deliver ITEBS may not be as straightforward as they presently seem.

4. Fiji's Potential in the Development of the IT-enabled Service Industries

Taking the above research findings into consideration, how does the Republic of Fiji Islands fare as a potential locale for placing outsourced IT-enabled business services?

Fiji is a small island economy in the South Pacific that promotes itself to international tourists as a tropical paradise of beautiful sun-drenched islands with white sandy beaches and swaying coconut trees, and with some of the best coral reefs and marine life in the world. It has population of 790,000 and it comprises 332 islands with a land area of 18,333 sq km. The current GDP per capita is about US$1200 per annum, and the country is characterized by a high level of food security (Asian Development Bank, 2000). Fiji's economy is dominated by the services sector that accounts for 70% of employment and income. The sugar industry is Fiji's single most important export earner, followed closely by tourism (*ibid.*). The emergence of a substantial garment industry in the 1990s, and more recently a mineral water industry, have helped diversify the economy and have proved the economy's ability to successfully nurture the development of new sectors. In the context of outsourcing ITEBS, it is also worth noting that Fiji is located on a time zone 12 hours ahead of GMT, thus making the country's location ideal for "overnight" processing of data sent from Europe and North America.

Below we assess Fiji's potential for becoming a successful exporter of outsourced and insourced ITEBS. In accordance with the findings of the preceding section, we focus first on labour costs, linguistic ability and the educational levels of the Fiji population, as well as the cost and quality of the national telecommunications infra-

structure, as the most important factors conditioning the development of ITEBS industries. Then we assess political and business-regulatory environment of foreign investment in general, and ITEBS specific regulatory environment in particular.

The lower wage cost is the major factor that makes Fiji an attractive choice for
ITEBS firms. Table 13.1 presents comparative wage rates for semi-skilled IT workers
in 5 countries, including Fiji. Two of them, Australia and New Zealand, are primary
target markets for outsourcing ITEBS from Fiji, and the remaining three, Singapore,
Fiji and India, can be considered as competing providers of these outsourced services.
Fiji's wage rates are around one-fifth of those in Australia and New Zealand. Both
countries' IT-services firms can obtain substantial cost reductions by locating their
services in Fiji. At the same time, Fiji's wage rates are comparable to those of its main
competitor—India—that has been successful in developing ITEBS exports.

Another key condition of attracting outsourced ITEBS pertains to the English language capabilities, education levels and requisite IT skills of the work force. Fiji has
both strengths and weaknesses in these areas. It has an English speaking, generally
well-educated population, but at the same time it suffers from the lack of adequate
IT-skills among students and graduates.

Table 13.1
Comparative Wage Rates for Semi-Skilled IT Workers

Country	F$ per hour
Australia	15–20
New Zealand	10–15
Singapore	4–8
Fiji	2–4
India	2–3

Source: TARPnz Strategic Methods Limited, 2001, p. 13.

English has become the official language in Fiji for state transactions and inter-
communal exchange, as well as for business. This is in spite of the fact that the 1997
Constitution recognizes that Fiji is a multilingual state with the main languages (Fijian, Hindi and English) being equal in terms of status, use and function. The reality
is that in a multi-ethnic Fiji, there is a need for a *lingua franca* and this need is per-
fectly filled by English (Fiji Islands Education Commission, 2000). English is also the

language of education used at all the three levels of education—primary, secondary and tertiary[34].

Fiji ranks favourably among its main potential competitors in ITEBS industries in terms of literacy levels of its population. Fiji's literacy rate of about 93% is higher than that of India (57%), China (84%), Dominican Republic (84%) and Mexico (91%), and is only slightly lower than that of the Philippines (95%) (UNESCO, 2002). The country also ranks favourably when the education index, prepared for the UNDP Human Development Report, is used as a measure of educational attainments[35]. Of the 24 small nation states included in the UNDP Human Development Report (UNDP, 1999), Fiji's education index placed the country at the top of the group (6th place). Overall, the population of Fiji achieved an educational attainment index of 0.88, which is higher than the average for the developing countries as a whole, and higher than the indices for South East Asia (0.67) and the Pacific (0.8). Based on the above figures, one can conclude that Fiji is better positioned than most of its main competitors in terms of the availability of well-educated employees required by ITEBS firms.

Although general education levels of Fiji's work force may be adequate, for IT-enabled business services such general levels, although indispensable, are often not sufficient. What is increasingly sought by IT firms is a computer-literate work force. In developed countries, major efforts are being made to fully computerize teaching and administration in secondary schools and to introduce computers extensively at primary education level. In developing countries, on the other hand, such efforts are rare, although there are noticeable exceptions. In Malaysia, for example, the "Smart Schools" initiative is aimed directly at producing a highly computer-literate generation of school leavers during this decade (TARPnz Strategic Methods Limited, 2001). In Fiji, such initiatives are lacking, mostly due to a shortage of funds, equipment, qualified teaching staff, and materials. As a result, very few secondary school leavers are computer literate and therefore only a small minority of secondary school graduates are expected to attain tertiary-level computing-related qualifications.

The quality and costs of telecommunication services is the third most important factor conditioning the development of ITEBS in low-income countries. In this area, Fiji has major weaknesses. A Discussion Paper published by the Pacific Islands Forum Secretariat in 2000 (Pacific Islands Forum Secretariat, 2000) points to high cost of Internet access in Fiji, especially for high volume (business) users, as compared to the Internet costs in developed countries. Although the situation has improved since 2000, costs of Internet access in Fiji are still considerably higher than in neighbouring developed countries; they are almost three times as high as the equivalent access costs

---

[34]   English is formally used as the instruction language from the fourth year of the primary school on. However, many primary schools use it as the instruction language from year one (Fiji Islands Education Commission, 2000).

[35]   The index is calculated on the basis of a country's combined primary, secondary and tertiary education enrolment levels along with its literacy rates.

in New Zealand and nearly two times as high as in Australia (ITU, 2003). However, in comparison to other Pacific Island countries, for which ITU has recently collected data, Fiji's rates are not high (see Table 13.2). Also, Fiji has some of the better tele-communications infrastructure and educational facilities in the region.

Table 13.2
Summary of Prices for Internet Access in
Selected Countries of the Pacific August 2003

Country	Total Internet Access Price Including Telephone Usage Charge, 20 Hours of Use (US$)
Fiji	31.74
French Polynesia	69.29
Marshall Islands	20.00
New Caledonia	80.34
Papua New Guinea	20.00
Samoa	42.97
Solomon Islands	91.15
Tonga	45.45
Vanuatu	46.70
Average	49.74

Source: ITU (2003).

Internet access costs in Fiji could be lower if not for the paternalistic and statist approach on the part of the Fiji Government to the management of the telecommunications sector. The Government protects the monopoly position of telecommunication service providers in spite of the lack of any credible international evidence that monopoly firms are capable of providing such services in an efficient and competitive way. As a result, Fiji has only one Internet service provider, Telecom Fiji[36].

On the positive side, a mention should be made of the high bandwidth capacity of the Southern Cross fibre-optic cable, which was launched in November 2000. The cable, linking Fiji directly to Australia, New Zealand and the U.S., has given Fiji the fast and reliable connection to the Internet that ITEBS need. In practical terms, it provides the carrying capacity for much increased level of telecommunications traffic,

[36] Five new licenses have allegedly been granted to ISPs; however, they will all be dependent on one company, FINTEL, for providing access to Internet connectivity.

needed, for example, for multiple call centres and high volume data transfers. The Southern Cross cable places Fiji, at least potentially, on a par with any other competing location globally. However, its full potential and impact is yet to be utilized through building the associated in-country infrastructure, enabling access to the cable capacity from various parts of the country. The most urgent (and already underway) is the laying of a fibre-optic cable connecting the Southern Cross cable's Suva exchange to the western districts of Viti Levu, where most ITEBS are expected to be located.

Generally speaking, Fiji has a market friendly business environment, sound macroeconomic policies, low inflation, and a relatively stable foreign exchange rate. It is, for example, noteworthy that in spite of the attempted coup in May 2000 and the subsequent political crisis and economic difficulties stemming from the trade embargo imposed by Fiji's major trading partners, the Fiji Government has managed to maintain the value of the Fiji dollar *vis-à-vis* major currencies.

Fiji also has a modern set of commercial contract laws that are enforced by the judiciary. It offers safety and security of persons and property, and enforcement of commercial contracts. At the same time, Fiji is notorious for ethnic tensions between its two major ethnic groups—Fijians and Indo-Fijians—that from time to time culminate in the form of major political crises. As a result, the country has had two effective and one attempted *coups d'état* in its recent history. However, after the general elections held in August 2001, the country has enjoyed democracy and political stability.

In terms of foreign investment, Fiji has mainly attracted investment from family owned companies from Australia and New Zealand. A recent survey by the Fiji Islands Trade and Investment Bureau (FTIB) has found that the majority of investors over the last decade have been owners/managers who are seeking a relaxed island lifestyle in an unpolluted, healthy environment with good quality schools, hospitals and a modern regional university with satellite network to 12 countries.

In its Web page, FTIB advertises the following advantages that Fiji offers to potential investors:

- Easy repatriation of capital and profits.
- An adaptable, productive, industrially disciplined and English speaking labour force with low wage rates.
- An attractive package of financial and other incentive schemes including a 13-year tax holiday and total freedom from import duties.
- Reasonable air and sea links with overseas markets
- Sophisticated telecommunication links with the rest of the world
- A well developed infrastructure, including electricity, water supplies and internal communications
- Availability of factory land and buildings at reasonable rates.

- Well-developed banking and financing institutions providing full ongoing financial services.

- Under the Foreign Investment Act 1999, the FTIB issues business certificates to all new proposals within 15 days of receiving complete proposals.

In spite of this encouraging advertisement, foreign investors' impression of the real situation is far from being rosy. Notably, "red tape" and bureaucracy are rated highly among obstacles to doing business in Fiji. Even if the FTIB issues a Foreign Investment Certificate promptly, a foreign investor is subsequently faced with a daunting task of obtaining numerous approvals from various authorities. Some of these approvals may take months or years to obtain. For example, it is reported that it typically takes about a year to obtain approvals from the Lands and Survey Department; some potential investors are reported to have gone bankrupt because of the delay (Asian Development Bank, 2000). Other obstacles pointed to by foreign investors include restrictive immigration requirements and procedures, a lack of clear accountability, responsibility and urgency among government agencies, and a low level of competence and efficiency among the staff handling foreign investment cases (TARPnz Strategic Methods Limited, 2001).

In order to facilitate the development and smooth functioning of ITEBS, the host country needs to put in place a regulatory framework that is specific to these industries, including various E-Commerce laws or "cyber-laws" (cf. GIPI, 2002). Fiji does not currently have a detailed legislative framework facilitating a smooth operation and protection of ITEBS. Notably, there is a lack of laws protecting personal privacy in the use of information, which means that client or competitive information may be at risk of unauthorized use or distribution. This will undoubtedly constitute a major deterrent, as many ITEBS deal with sensitive information. Legislation of E-Commerce is underway, but the proposed "White Paper on Electronic Commerce" is criticized as being inadequate in regulating such specific issues as electronic filling of documents, recognizing electronic records for subsequent references, accepting electronic documents as legally binding, and the requirements for using a secure electronic signature (TARPnz Strategic Methods Limited, 2001). The Government's approach seems to be that the initial ITEBS establishing in Fiji will generate a need to regulate specific E-Commerce issues.

Nevertheless, the Fiji Government has recognized the potential for the development of the IT-enabled business services and is keen to quickly establish a niche in this rapidly expanding market. The Government considers that Australia and New Zealand are most likely to be the main markets of ITEBS because of their closeness to Fiji. FTIB has already approved four major projects in this industry; however, the implementation of these projects has been delayed due to the lack of confidence on the part of the investors. Fiji simply does not have an established reputation as a destination for ITEBS investments. One of these projects involves the construction of a 105 seat, 24-hour call centre to service overseas clients, as a joint venture between the Fiji International Telecommunications Limited (FINTEL) and an overseas company. The joint venture will employ 400 staff. Another 500 seat call centre is being considered by a large foreign investor[37]. Apart from call centres, a project that has attracted a lot of attention in Fiji is the proposed development of an audio-visual industry, for which the Fiji Audio Visual Commission (FAVC) has been established. As a starter, the "Studio City Zone" has been set up in Yaqara, on the northern shores of the main island of Viti Levu. The 2,200-hectare Studio City is a tax free zone for business and individuals. It is designed to attract investors in filmmaking, tourist resorts, residential housing, retail, and audio-visual education fields (FAVC, undated).

4.1 Recent Fiji Back Office Developments

Since 2001 the Fiji Islands Trade and Investment Bureau (FTIB) has been lobbying the Fiji Government to allocate resources to establish an information technology park and also to fund a more aggressive targeted marketing campaign to establish Fiji as a new location for back offices IT services.

By November 2003 the results of a modestly funded FTIB marketing effort are bearing fruit. Several pioneering firms are now in operation in the banking and credit card services, insurance claims processing and airline industry voucher processing, as well as call centres for IT support services.

Quest Ltd, a subsidiary of the ANZ Bank Limited, has established an IT business centre in Suva that employed 53 full-time staff by October 2003 with expansion plans to double its employment during the coming year. It provides 24-hour on-line technical support services to 22 countries on IT systems support through its call centre in Suva. It also provides ANZ Visa Card support services and back office corporate services in finance and marketing to ANZ banks located in several of the other Pacific Island countries.

Affiliated Computer Services opened its doors on 29 October 2003, employing 60 trained staff to undertake airline voucher processing for Air New Zealand. It has rented two floors in the new Fijian Holding Limited office tower in central Suva. It also plans to expand rapidly to increase its workforce to over 120 persons within a year of commencing operations.

Computech Limited has also commenced operations in the software development industry employing more than a dozen IT professionals including skilled programmers with university IT degrees. It has a contract with a USA state government for software development work.

Other developments include Colonial Insurance that undertakes medical claims processing for Pacific Island clients, Telecom Fiji that operates a 29-seat call centre and the Westpac Bank that services its regional bank office IT network from its Suva-based IT operation.

---

[37]   Interview with Mr Ledua, the CEO of FINTEL, 4 March 2002.

5. Policy Recommendations

FTIB has found that encouraging the first few firms to enter this new industry has been slower than expected because of the lack of agglomeration economies for the first firms. Now there is a group of pioneers in the key IT industries of insurance, banking and finance and airline industry, and FTIB is confident that other firms will be attracted to Fiji. However, if this good start is to accelerate, the Fiji Government must allocate more resources to an international marketing campaign by FTIB to sell Fiji as a sound location for investment in the IT-enabled business service industry.

To attract IT-enabled business services, the Government must be willing to consider and implement a range of policies specifically designed to create a regulatory environment that is conducive to the development of such services and to promote Fiji as an efficient, technology-driven and "wired" economy.

We recommend that the Government quickly put in place the appropriate policies, institutions, regulatory and legal framework to nurture the development of this new industry opportunity for Fiji. Our main recommendations are derived from our analysis of Fiji's potential for the development of ITEBS industries. These are as follows:

5.1 Launch a National Strategic Plan and Policy Framework for ITEBS Industry Development

The Fiji Government recognizes the urgency in implementing strategies aimed at propelling the nation into the digital broadband era. It has identified IT as a potential growth sector and is currently developing a policy framework to provide guidelines for the development of the industry (Minister of Finance and National Planning, 2001). The Government knows that many other countries have already established a substantial competitive advantage in the industry and that an ITEBS industry is unlikely to develop spontaneously in Fiji without a major effort from the Government. We recommend that the Government play a major role in promoting Fiji as a profitable location for the industry. Political stability and a soundly managed economy are top of the long list of conditions necessary for successfully launching this new industry.

We recommend that the Fiji Government develop a strategic plan for the industry that clearly defines the Government's role, vision, mission and specific timetable of activities to put in place the legislation, institutions and policy measures. This approach has proved valuable in other countries such as Australia and Malaysia (Ministry of Communications, Information Technology and the Arts, 1998; and MultiMedia Development Corporation, 1997).

The strategic plan should identify resources and institutions responsible for implementing the plan and the new policy measures to assist innovative small and medium sized enterprises hurdle the early obstacles and to assist fledgling local technology companies gain the management skills, marketing know-how and knowledge to safeguard their intellectual property. Fiji should adopt a proactive, investor-focused approach to attracting capital and expertise. This approach would involve the Government identifying potential overseas investors who are evaluating alternative locations and inviting them to Fiji to personally assess the attractiveness of Fiji as a location for their investment.

5.2 Streamline the Investment and New Business Approval Processes by Implementing New Measures to Reduce the Time Lag for the Various Forms of Approvals Required by New Investors

As was pointed out in the preceding section, various bureaucratic hurdles faced by foreign investors during the investment approval process have a major deterrent effect on their propensity to invest in Fiji. The Government should give high priority to implementing new measures to shorten the time that it takes for investors and eliminate unnecessary hurdles to gaining the approvals necessary for establishing a new business enterprise.

We recommend setting up an office to drive the development of the new ITEBS industry and to act as an effective one-stop super shop for IT investors. The Fiji Government has a specialized agency, the Fiji Islands Trade and Investment Bureau (FTIB), whose sole function is the promotion of trade and investment. It aims to provide investors with a "one-stop-shop" service and guide them through the complex array of approvals required for establishing new enterprises. However, as we pointed out in the preceding section, the reality is that this "one-stop-shop" works as such only for the issuance of a Foreign Investment Certificate, while numerous other approvals must be obtained from other Government departments. It makes sense for the proposed office to facilitate the whole investment process and not only its first step. It also makes sense to locate this office within FTIB. FTIB has identified the ITEBS industries as the next new industry in Fiji and jointly commissioned consultants to prepare a marketing plan to guide them in promoting the industry.

5.3 Promote Fiji Aggressively as a Preferred Destination for IT-enabled Services Investments

Promoting Fiji to prospective IT investors should also be a major function of the proposed FTIB office. The promotion strategy should include activities such as:

- establishing a Fiji ITEBS Industry Web Site which provides information on investment opportunities;
- developing an international network of contacts of firms engaged in back office services and the establishment and maintenance of a data base; the data base should contain details of the leading firms in this industry;
- producing and distributing a high quality promotional material on ITEBS industries development opportunities;

- conducting international promotional seminars on ITEBS industries development to be held in Sydney, Australia, Auckland, New Zealand, Los Angeles, U.S. and in Europe;

- identifying land and office space in the Nadi/Lautoka area suitable for ITEBS industries operations;

- developing a skills training programme for ITEBS industries employees through the Fiji National Training Council;

- monitoring software and hardware development which supports back office and call centre development;

- monitoring the development of ITEBS industry developments world-wide and providing advice to government on its development strategy.

FTIB has experience in driving the development of a new industry. It played a major role in the establishment of the garment industry that developed rapidly in the 1990s, largely as a result of a new incentive package combined with preferential trade agreements with Australia and USA. To assist foreign investors to set up production quickly, FTIB established Kalabo industrial estate comprising standard factory buildings in the capital city, Suva. It rents factory space directly to investors thus saving them from the risks and delays associated with acquiring industrial land and constructing purpose-built factories.

5.4 Allocate Substantially More Resources to Support IT in Schools and Post-secondary IT-training Courses

Urgent action should be taken to bring IT exposure and training to secondary schools in Fiji. This will require a substantial commitment of funds for equipment and teacher training programmes. Such commitment is essential if Fiji is to make a meaningful attempt at developing a knowledge economy. It is also essential if Fiji intends to exploit the opportunities for participating in higher value ITEBS industries in the future. Fiji must address the seriously low level of access to IT in schools and homes through much more innovative approaches and strategies for capital investment in education and telecommunications services pricing. The Government, through FTIB, should also work with private training providers and ITEBS industry investors to establish preparatory training facilities and capabilities for this industry workers.

5.5 Enact Essential "Cyber-laws"

The Government should enact a number of essential cyber-laws such as a Digital Signature Act to facilitate E-Commerce by providing an avenue for legally recognized on-line transactions, a Computer Crimes Act to penalize various activities related to the misuse of computers and unauthorized access to computer material, and other "cyber-laws". Together with progressive Intellectual Property Laws, the proposed "cyber-laws" would not only facilitate the development of ITEBS industries in Fiji but they would also allow the country to become a regional leader in implementing these laws.

5.6 Allow Unrestricted Employment of Foreign Knowledge Workers

The current system for obtaining work permits is expensive, slow and cumbersome. It is restrictive and is a source of concern for foreign investors. We recommend that Fiji allow unrestricted employment of foreign knowledge workers to accelerate economic development.

5.7 Deregulate the Government Telecommunications Monopoly to Provide Firms with Globally Competitive Telecommunications Tariffs

International experience clearly demonstrates that competition in telecommunication leads to lower tariffs and improved services. The Fiji Government should deregulate the telecommunications industry and encourage competition. It should not continue its current incremental approach by trying to introduce a modest degree of competition within the existing ownership structure and the existing restrictions on foreign investment in the sector. Instead, it should take a more radical approach and open the telecommunications markets to both domestic and foreign competition. Many countries have done so over the last two decades and both the service quality and operational efficiency have improved significantly. Needless to say, the consumers, both business and individual, have benefited from better quality and lower prices of telecommunications services.

5.8 Kick-Start the New Industry through Joint-venture Investment in Call Centres by Telecom Fiji and FINTEL with Overseas Partners

Both FINTEL and Telecom Fiji have plans to set up call centres to service overseas and local clients. These projects will be pioneering enterprises and they will most likely be undertaken in partnership with foreign companies that are already well established in the industry. Attracting the first entrants into a new industry is always difficult because they may not achieve the agglomeration economies that are associated with a cluster of similar firms operating in the same location. We support the plans of Telecom and FINTEL to establish the first call centres in Fiji as a strategy for pioneering the start of a potential new industry. As was mentioned in the preceding section, FINTEL plans to construct a 105 seat call centre on its land in Suva at its earth station at Vatuwaqa. The Government has also been considering the TARPnz report recommendation that an IT business park be developed specifically for IT-enabled services, as a means of engendering and sustaining new performance standards in the industry. It is envisaged that the park could be owned and operated by a joint venture consortium involving Telecom Fiji, FINTEL and private sector investors (TARPnz Strategic Methods Limited, 2001).

5.9 Establish a Service Quality Assurance Scheme through a System of Supplier Accreditation

We recommend the establishment of a quality assurance scheme for the new industry. Such a scheme would involve the accreditation of new firms that join the industry and adherence by firms to various industry codes of conduct within the different services that provide an assurance of quality and consistency of service.

6. Conclusions

Advances in ICTs have been driving the expansion of the service sector over the last two decades and making many business services more and more tradable internationally. The diffusion of ICTs within the business sector has lowered the costs of transacting and communicating among firms, allowing specialized slivers of the value chain to be located in a widening range of locations far from the client base. This has led to global outsourcing to developing country suppliers of many IT-enabled business services by firms based in developed countries. The fastest growing areas lie in outsourcing of "back office" functions, the trend being driven mostly by the lower costs of critical HR inputs.

Migration of business services to low-income countries is driven primarily by the lower costs of labour. Developed countries' companies may save as much as 50% of their operational costs by shifting ITEBS to developing countries. Another critical factor in attracting ITEBS is the linguistic ability of a host country's population, as well as its education and skill levels. To compete in higher-value added segments of ITEBS, a pool of talented labour must be available. The national telecommunications infrastructure must offer acceptable levels of service in terms of cost and quality.

Meeting the above conditions is necessary but not sufficient for a country to be a preferred destination for outsourced or in-sourced ITEBS. Research points to other key enablers of competitiveness in the industry. One of them is the national regulatory environment that affects the development of tradable ITEBS, including a legal framework for electronic commerce. Another one is marketing and customer service—the service provider's ability to build its credibility by effectively promoting its services, establishing brand recognition and maintaining a high and consistent level of customer service.

Several developing countries have succeeded in attracting a substantial amount of outsourced ITEBS, and many others have the potential to follow suit. Fiji is one of the latter. This country's advantages include:

- Substantially lower wage cost structure than that of developed countries
- An English-speaking, comparatively well-educated population
- Market-friendly business environment
- Sound macroeconomic management and relatively stable currency
- Safety and security of persons and property
- Enforcement of commercial contracts
- High volume capacity telecommunication links with the world (through the Southern Cross fibre-optic cable)

- Location on a time zone 12 hours ahead of GMT
- Favourable FDI climate, allowing for easy repatriation of capital and profits

However, Fiji has also certain disadvantages, which it must overcome if it aspires to become a magnet for ITEBS industries. These include:

- A lack of strategy and policy framework for the development of ITES industries
- Overly bureaucratic and time consuming new business approval process
- Insufficient IT exposure and training of secondary school students
- A lack of "cyber-laws"
- Restrictive foreign workers employment regulations
- Protected telecommunications markets and the resultant high costs of services, including Internet access
- Only an infacy of successful ventures, too few (so far) to create agglomeration economies
- A lack of quality assurance of the services delivered by the sector
- Insufficient promotion of Fiji as an ITEBS investment destination
- Susceptibility to political instability

The favourable conditions Fiji possesses with respect to ITEBS can only be fully exploited if these disadvantages are eliminated or minimized. Therefore, we put forward a number of policy recommendations that address the above-listed disadvantages. The recommended measures include:

- Launching a national strategic plan and policy framework for ITEBS industry development
- Streamlining the investment and new business approval process
- Allocating more resources to support IT training in schools
- Enacting essential "cyber-laws"
- Allowing unrestricted employment of foreign knowledge workers
- Deregulating the telecommunications markets
- Kick-starting the new industry through joint ventures with overseas partners
- Establishing a service quality assurance scheme

We conclude that the country has considerable potential to become a magnet for IT-enabled business services outsourced and in-sourced by companies from specifically targeted developed regions. We recognize however that the ITEBS industries are unlikely to develop spontaneously in Fiji. The Government will need to play a major role in overcoming Fiji's disadvantages, in promoting Fiji as an ideal location for these industries, and in ensuring continued political stability.

References

Anonymous (2001a) Management Trends in Outsourcing, Outsource2India.com,
    www.outsource2india.com/why_outsource/articles/management_trends.asp

Anonymous (2001b) Back Office to the World, *The Economist*, **359**, 8220, 59. May 5th.

Apte, U. and Mason, R. O. (1995) Global Disaggregation of Information-intensive Services,
    Management Science, **41**, 7, 1250–1262.

Aron, R. and Singh, J. (2002) IT-enabled Strategic Outsourcing. Knowledge-Intensive Firms,
    Information Work and the Extended Organizational Form.
    knowledge.wharton.upenn.edu/PDFs/1071.pdf

Asian Development Bank (2000) *Republic of Fiji Islands 1999 Economic Report*. Manila: Philippines.

Burnetti, A., Kisunko, G. and Weder, B. (1997a) Credibility of Rules and Economic Growth: Evidence
    from a Worldwide Survey of the Private Sector, *Policy Research Working Paper* 1760;
    Washington, DC: World Bank.

Burnetti, A., Kisunko, G. and Weder, B. (1997b) Institutional Obstacles to Doing Business. Region by
    Region Results from a Worldwide Survey of the Private Sector, *Policy Research Working Paper*
    1759; Washington, DC: World Bank.

Carmel, E. and Agarwal, R. (2000) Offshore Sourcing of Information Technology Work by America's
    Largest Firms. *Technical Report*, Kogod School, American University, Washington D.C..

Casson, M., and Wadeson, N. (1998) Communication Costs and the Boundaries of the Firm,
    International Journal of the Economics of Business, **5**, 1, 5–27.

Corbett, M. A. and Associates (1999) Outsourcing Set to Explode in Asia,
    www.firmbuilder.com/articles/19/42/407/.

Corbett, M. A. and Associates (2001a) Outsourcing's Global Marketplace,
    www.firmbuilder.com/articles/19/42/408/.

Corbett, M. A. and Associates (2001b) Taking the Pulse of Outsourcing—Data and Analysis from the
    2001 Outsourcing World Summit, www.firmbuilder.com/articles/19/42/676/.

CyberAtlas (2003) BPO Market to Reach $122B in 2003, www.internetnews.com/stats/article.php/2220371.

Davis, C. H., McMaster, J. and Nowak, J. (2002) IT-Enabled Services as Development Drivers in Low-
    Income Countries: The Case of Fiji, *Electronic Journal on Information Systems in Developing
    Countries*, **9**, 4, 1–18.

Duncan, R., Suthbertson, S. and Bosworth, M. (1999) *Pursuing Economic Reform in the Pacific*. Asian
    Development Bank.

Dunn and Bradstreet (2000) Barometer of Global Outsourcing—The Millennium Outlook.
    www.dnbcollections.com/.

Elsham, R. (2001) GE Capital Shows India's IT-enabled Service Might. Yahoo Finance
    biz.yahoo.com/rf/011204/bom94182_4.html, December 4.

Evans, P. and Wurster, T. H. (2000) *Blown to Bits: How the New Economics of Information Transforms
    Strategy*, Cambridge, Mass: Harvard Business School Press.

FAVC (undated) *An Era of Great Opportunities*. Fiji Audio Visual Commission. Promotional Brochure.
    Suva, Fiji Islands.

Fiji Islands Education Commission (2000) *Learning Together: Directions for Education in the Fiji Islands*. Government of Fiji (Ministry of Education), November.

GIPI (2002) *The Keys to the Back Office: Building a Legal and Policy Framework to Attract IT-Enabled Outsourcing*. Washington, D.C.: Global Internet Policy Initiative.

Hyder, E., Kumar, B., Mahendra, V., Siegel, J. Gupta, R. Mahaboob, H. and Subramanian, P. (2001) The *Capability Model for IT-Enabled Service Providers*. Vol. 1: *Overview*. Vol. 2: *Practice Descriptions*. Pittsburgh: School of Computer Science, Carnegie Mellon University.

ITU (2003) *World Telecommunication Development Report, 2003*. International Telecommunication Union. www.itu.int/ITU-D/ict/publications/wtdr_03/index.html.

Kobrin, S. J. (1999) Development After Industrialization: Poor Countries in an Electronically Integrated Global Economy, In Hood, N. and Young, S. (eds.) *The Globalization of Multinational Enterprise Activity and Economic Development*. Chapter 6. London: Macmillan.

McCracken, J. (2003) Canada, the High Touch/High Quality Call Center Outsourcing Option, *BPO Outsourcing Journal*, October. www.bpo-outsourcing-journal.com/issues/oct2003/canada.html

McMaster, J. and McGregor, A. (1999) *The Fiji Service Sector: Opportunities for Growth*. A Consulting Report Prepared for the Ministry of Commerce, Fiji Government. Suva: Fiji Islands.

Minister of Finance and National Planning (2001) *2002 Government Budget Address*. Suva: Fiji Islands, November.

Ministry of Communications, Information Technology and the Arts (1998) *A Strategic Framework for the Information Economy: Identifying Priorities for Action*. Canberra: Australia.

Miozzo, M. and Soete, L. (2001) Internationalization of Services: a Technological Perspective, *Technological Forecasting and Social Change*, **67**, 2&3, 159–185.

MultiMedia Development Corporation (1997) *Multimedia Super Corridor: Unlocking the Full Potential of the Information Age*. Malaysia: Cyberjaya.

NASSCOM-McKinsey (1999) Highlights of the NASSCOM-McKinsey Study Report 1999. National Association of Software and Service Companies (India), December. wwwca.customerasset.com/papers/Outsourcing%20to%20India%20-%20A%20NASSCOM-%20McKinsey%20study.pdf

Pacific Islands Forum Secretariat (2000) *Internet Access, Prices and Development*. Pacific Islands Forum Secretariat Discussion Paper 20, Suva, Fiji Islands, October.

Primo Braga, C. (1996) The Impact of the Internationalization of Services on Developing Countries, *Finance and Development*, **33**, 1, 34–37.

Quinn, J. B. (1992) *Intelligent Enterprise*. New York: Free Press.

Quinn, J. B. (2000) Outsourcing Innovation: the New Engine of Growth, *Sloan Management Review*, **41**, 4, 13–28.

Riddle, D. (2000) Export Development in the Digital Economy. A Focus on Business and Professional Services, www.intracen.org/execforum/docs/ef2000/db2rdl.htm.

Rundell, J. (2003) Offshore Outsourcing, Presentation at *Sourcing Interest Group—Australian Regional Meeting*, 28 August.

Serven, L. (1997) Uncertainty, Instability, and Irreversible Investment: Theory, Evidence, and Lessons from Africa. *Policy Research Working Paper* 1722.Washington, DC: World Bank.

TARPnz Strategic Methods Limited (2001) *A Policy Framework for Developing IT Enabled Industries in Fiji*. A Consulting Report for the Fiji Government.

UNCTAD (1998) Scope for Expanding Exports of Developing Countries in Specific Services Sectors Through All GATS Modes of Supply, Taking Into Account Their Interrelationship, the Role of Information Technology and of New Business Practices. United Nations Conference on Trade and Development, note by Secretariat, document TD/B/COM.1/21, 24 July.

UNCTAD (2002) *E-Commerce and Development. Report 2002*. New York: United Nations Conference on Trade and Development.

UNCTAD (2003) Business Process "Outsourcing Services for Economic Development", Chapter 5 in *E-Commerce and Development. Report 2003*. New York and Geneva: United Nations Conference on Trade and Development, document UNCTAD/SIDTE/ECB/2003/1.

UNDP (1999) *Human Development Report*. United Nations Development Program.

UNESCO (2002) Indicators on literacy. United Nations Educational, Scientific and Cultural Organization www.un.org/Depts/unsd/social/literacy.htm.

Wymbs, C. (2000) How E-Commerce is Transforming and Internationalizing Service Industries, *Journal of Services Marketing,* **14**, 6, 463–478.

Appendix
Types of IT-enabled Business Services

Skill Level and Knowledge Intensity	Administrative	Customer Service	Technical
Low	Data entry; clerical	Call centre; routine queries; order taking; direct mail order processing	Transcription; indexing and abstracting
Intermediate	Secretarial; data capture and processing; mailing lists; credit card application processing	Account queries; after sales support; insurance claim processing, processing of warranty card and claims	Website design and management; medical records management; medical transcription
High	Accounting; payroll; electronic publishing; facilities management; management consultancy; legal services	Problem and dispute resolution	Software development; R&D; application hosting; technical writing; computer aided design; tele-medicine; engineering design; education; animation

Source: Adapted from McMaster and McGregor (1999)

About the Authors

Dr. Jan Nowak is Professor of Business and Co-ordinator of the MBA Programme at the University of the South Pacific (USP) in Fiji. He specializes in marketing and international business, the areas in which he teaches, conducts research, consults with companies, and develops and facilitates professional training courses. He is also Adjunct Professor at the Faculty of Business, University of New Brunswick in Saint John, NB, Canada. Professor Nowak commenced his professional career in Poland, where he received his Master of Management and Ph.D. in Business Administration degrees from Warsaw University. In 1990, he moved to Canada and joined the University of New Brunswick in Saint John (UNBSJ). While on the Faculty of Business, UNBSJ, he taught at both undergraduate and graduate levels, and directed an MBA programme and a management training centre. At the same time, he participated in several international projects sponsored by CIDA and travelled extensively to Asia and Central Europe. In 2001, he took up a new position at the University of the South Pacific in the Fiji Islands. Dr. Nowak has published five books, more than 25 papers in international refereed periodicals and conference proceedings, and numerous reports, working papers, book contributions and other research outputs. His articles have appeared in such academic journals as the *Journal of Transnational Management Development, The Electronic Journal on Information Systems in Developing Countries, Journal of Euromarketing, Journal of East-West Business* and *Journal of International Food and Agribusiness Marketing.*

Professor Jim McMaster is an economist who joined USP in 1999 as Director of the Pacific Institute of Management and Development. He teaches South Pacific Business Environment and Business Economics on the MBA. Before joining USP he was Dean of the Faculty of Management at the University of Canberra where he was previously the Director of the MBA programme. He has undertaken many consulting assignments for the major international organizations, including: the World Bank in Kazakhstan, Malaysia, Korea, Indonesia, Philippines, and Singapore; Asian Development Bank in Malaysia and Philippines; International Monetary Fund; United Nations Development Programme in Vietnam, Papua New Guinea, Fiji; Forum Secretariat in the South Pacific; Commonwealth Secretariat; East-West Centre in Hawaii, Tonga, Vanuatu, Fiji, Samoa; Asian and Pacific Development Centre in Kuala Lumpur and the National Centre for Development Studies at the Australian National University, Canberra.

Dr. Charles Davis is Professor in the Faculty of Communication and Design at Ryerson University in Toronto, Canada, and holder of the Edward S. Rogers Sr. Research Chair in Media Management and Entrepreneurship. He was formerly in the

Faculty of Business at the University of New Brunswick, Saint John, where he taught graduate and undergraduate courses in Foundations of e-Business, Strategy, Enterprise Integration, Customer Relationship Management, and New Products Management. His current research focuses on innovation in the media industry, regional innovation clusters, and development of proficiencies in software-enabled work environments. He earned his Ph.D. in science and technology policy and management from the Université de Montréal.

14. The Provision of Internet Services in India

Peter Wolcott

Department of Information Systems and Quantitative Analysis
College of Information Science and Technology
University of Nebraska at Omaha
pwolcott@mail.unomaha.edu

Abstract

Critical to the development of the Internet in any country is the development of the organizational infrastructure necessary to bring Internet services to the populace. This descriptive study examines the development of the market for Internet services in India. As a result of key policy decisions during the 1990s, India experienced a dramatic increase in the number of Internet users, Internet Service Providers (ISPs), and ISP points of presence from 1998 onward. An important element of this growth was the creative tension between policy-makers, the incumbent state-owned telecommunications service providers, and the new ISPs. The competitive landscape in India is examined through the lense of a framework for analyzing the strategies of ISPs in Europe. While this framework partially explains the evolution of Internet services in India, it is incomplete in several respects, failing to account for distinctive characteristics of Indian ISPs, and key government policies of gradual liberalization of communications and universal service provision. India's experience offers a number of insights into and lessons for the development of the Internet. The country's experience has been, on the whole, successful, and offers one model of how countries can transition from Internet laggard to one with a vibrant Internet presence.

1. Introduction

In the years following its founding in 1947, the Republic of India implemented a set of policies designed to enhance the self-reliance and political integrity of this highly diverse and expansive country. A highly regulated and protectionist domestic market emerged, overseen by a bloated bureaucracy. In response to economic crises during the 1970s and 1980s, domestic and international pressure, and global trends in trade and investment, the Government of India implemented a set of economic reforms in the 1990s. The realization that information and computing technologies (ICT) constitute a critical part of the economy of modern states was an important element of these reforms.

In 1998, Prime Minister Vajpayee offered a vision of a 21st century India in which information technology (IT) would enable significant change in the lives of individuals, communities and governments, and in India's relationship with the world. In his words (Vajpayee, 1998),

> [IT] is revolutionizing life on this planet like no other technology has in human history. It has been impacting on the economy, communication, culture, educational system and social interaction in all the countries, bringing them closer in a world transformed into a Global Village and laying the foundation for a new civilization. India, as the cradle of civilization, is poised to become a major IT power in the coming years and contribute to the realization of its many promises for our own benefit and for the global good.

The Internet is central to the new vision of India as an IT power in many respects. Technologically, the Internet provides the infrastructure for a range of applications supporting person-to-person, business-to-consumer, and business-to-business interaction. In many nations, the desire to expand the presence of the Internet has driven the build-up of extensive communications infrastructure. Economically, the Internet supports or enhances critical technology industries such as telecommunications and export-oriented software development, creates opportunities for innovation and new business, offers new channels by which companies can reach their customers, enables the growth of domestic and international markets, and facilitates the execution of transactions even when the parties are geographically distant. Politically, the Internet may support more direct and transparent relationships between governments and the populace. In addition, the Internet has been a driver of changes in policies, laws and government services to accommodate the digital age. Socially, the Internet, in its very nature, is a technology for the masses; Internet penetration rates are often viewed as an indicator of a nation's technological status.

Critical to the development of the Internet in any country is the development of the organizational infrastructure necessary to bring Internet services to the populace (Wolcott *et al.*, 2001). A key question is how India, with extensive poverty and a large and cumbersome bureaucracy, can transform itself into a nimble provider of Internet services. A solid understanding of the factors that promote or inhibit such development can help policy-makers establish an environment in which the Internet can thrive. This descriptive study examines the development of the market for Internet services in India. In particular, it reveals the creative tension between policy-makers, the incumbent state-owned telecommunications services providers, and the new Internet Service Providers (ISP) that has fuelled a good deal of the growth of the Internet observed in India (Figure 14.1). Developments in India can partly be understood by applying a framework for analyzing strategies of ISPs (Wierstra *et al.*, 2001).

Figure 14.1
Key Relationships Shaping the Growth of the Internet

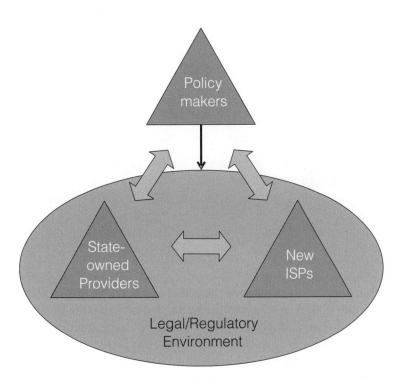

Following this introduction, the second section provides some background on the evolution of the telecommunications regime in India. The third section describes the birth of the Internet in India through 1998. Section four describes the growth of Internet services since 1998 policy changes permitting commercial ISPs to proliferate. In the fifth section, a framework for analyzing strategies of ISPs is used to understand developments in India. Throughout, the role of policy-makers, private ISPs and state-owned providers is discussed. Portions of this chapter are based on Wolcott and Goodman (2003).

2. Background

As a result of the Telegraph Act of 1885, the Government of India held a total monopoly on all types of communications until the late 1990s. The government controlled not only the operational aspects of telecommunications services, but also the policy-making and regulatory functions (Dossani and Manikutty, 2002).

In the mid-1980s, the first tentative steps towards a more liberalized regime were taken. Two state-owned corporations were founded to provide limited communications services—Mahanagar Telephone Nigam Limited (MTNL) for telecommunications services in Mumbai and Delhi and Videsh Sanchar Nigam Limited (VSNL) for international long-distance. The Department of Telecommunications (DoT) and its parent ministry, the Ministry of Communications (now called the Ministry of Information Technology) handled all other aspects of telecommunications in India, including licensing, policy-making, regulation, tariff setting and cross-subsidies. The *National Telecommunications Policy, 1994* (NTP 1994) sought to expand telecommunications services throughout the country, in part by opening the possibility of the private sector providing some basic services.

To fulfil commitments made when India joined the World Trade Organization (WTO) in 1995, the Ministry of Communications separated its regulatory and operating functions. As a result, the Telecommunications Regulatory Authority of India (TRAI) was created in 1997. Operating functions remained in the DoT. The DoT divisions providing operational services were transformed into a corporation—Bharat Sanchar Nigam Limited (BSNL)—in 2000, and VSNL was successfully privatized in 2002.

In liberalizing the telecommunications regime, the Government of India has tried to balance competing concerns. On the one hand, policy has been driven by a desire to allow market factors to shape the development of telecommunications in general and the Internet in particular. On the other hand, successful privatization of state-owned entities requires that their income streams be sustained to keep them attractive to potential investors. Stripping them of all their monopoly rights at a stroke would likely have devastated these entities and greatly reduced their market value. The Indian government's solution to this dilemma has been to peel away the monopoly rights of VSNL and BSNL gradually, but persistently, forcing them to compete with new entrants, while leaving them with the fiscal health to do so.

Table 14.1 shows the years in which monopoly rights were removed from the state-owned providers.

Table 14.1
Loss of Monopoly Rights

Monopoly Right	Monopoly Holder	Year Monopoly Ended
Internet service provision	VSNL	1998
International gateways	VSNL	1999
Domestic long distance	BSNL	1999
Access to international terrestrial cable	VSNL	2000
Domestic local telephone	MSNL, BSNL	2002
International telephony	VSNL	2002
International long distance	VSNL	2002

3. Birth of the Internet in India (1986–98)

From 1986–98, the Indian government supported a variety of programs to establish nationwide networks. Targeting the needs of the government, academia and industry, these projects contributed to the creation of organizations and institutional know-how that helped support the Internet in its early days. Three of these were initiated in 1986–87—INDONET, initially to serve the country's hundreds of IBM mainframe installations; NICNET, the National Informatics Centre Network, a nationwide very small aperture terminal (VSAT) network for public sector organizations; and the Education and Research Network (ERNET), to serve the academic and research communities. VSNL also initiated several wide-area networks (WAN) during the 1980s.

ERNET (www.eis.ernet.in) was established in 1986 by the Department of Education (DoE) and seven other government organizations—the National Centre for Software Technology (NCST); the Indian Institute of Science (IISc), Bangalore; and the Indian Institutes of Technology (IITs) in Chennai (Madras), Kanpur, Kharagpur, Mumbai and New Delhi. The project received technical and financial support from the UN Development Program (UNDP). The goals of the project were to set up a nationwide computer network to enable the academic and research communities to conduct research and development in computer networking, and to provide network training and consulting services (Anonymous, 1998a). The NCST was the first institution in India to establish an international connection to the Internet and subsequently acquired the responsibility for managing the *.in* (India) national top-level domain, since the organization established a root-server and the first *.in* Internet domain (ncst.ernet.in).

The first connection to the global Internet, a 9.6 Kbps link between NCST and UUNet (UNIX-to-UNIX Network) Technologies in the United States, was established in 1989. The international link was up-graded to 64 Kbps in 1992.

In 1995, VSNL became the first commercial Internet service provider by offering public Internet services via a gateway earth station and router in Mumbai that provided a single connection to MCI in the USA. Local access nodes were installed in Kolkata, Chennai, Mumbai and New Delhi, permitting connections via dial-up lines through the DoT or MTNL or an I-NET connection. Within two years, VSNL/DoT had 75,000 Internet subscribers on the network and double that number in 1998 (Anonymous, 1998b; Kumar, 1998; Pai, 1998).

4. Internet Boom

In the latter half of the 1990s, the pace of reform quickened considerably, with the Internet playing an integral role, both as a factor driving reform and as a beneficiary of change. The election of the Bharatiya Janata Party (BJP) in 1997 signalled enhanced interest in IT and the Internet. The BJP advocated economic liberalization and listed IT as one of the government's top five priorities, along with more traditional issues such as the provision of potable drinking water and education.

One of the clearest signs of a changing government attitude was the creation of the National Task Force on Information Technology and Software Development (it-taskforce.nic.in/). The National Agenda for Governance, issued in March 1998 by the BJP, outlined India's goal of becoming a software superpower and announced the creation of a National Informatics Policy aimed at achieving this end. The Task Force was directed to produce an Action Plan within 30 days and a National Informatics Policy within 90 days. The Task Force was created on 22 May 1998 under the chairmanship of Jaswant Singh, Deputy Chairman of the Planning Commission.

Less than three weeks after convening, the Task Force published a Background Report on the state of IT in India. An Action Plan, outlining 108 recommendations, followed on 4 July 1998 (Anonymous, 1998c). The Task Force could act quickly because it built upon the experience and frustrations of state governments, central government agencies, universities, the software industry, other nations, and the recommendations of the WTO, the ITU and the World Bank.

The Internet was featured prominently in the Action Plan, with nine of the recommendations, including the very first, specifically addressing Internet development issues. The first recommendation directed the DoT "and authorized ISPs" to open Internet access nodes in all district headquarters by 26 January 2000.

The IT Action Plan of 1998 was followed by several major pieces of legislation changing the telecommunications regime. The *New Telecommunications Policy, 1999* (NTP 1999) went much further than NTP 1994 in liberalizing the telecommunications sector (Anonymous, 1999e). The *Information Technology Act 2000* placed electronic transactions and commerce on a firmer legal footing (Anonymous, 2000). The pending Convergence Bill or an alternative unified license policy seeks to establish an integrated regulatory regime for voice, data and broadcasting communications (Rambabu, 2004).

The *New Internet Policy* (Anonymous, 1997) went into effect in November 1998, allowing any Indian company with a maximum foreign equity of 49% the possibility of providing Internet services. It established three categories of license:

- Category "A" licenses apply to the entire country.
- Category "B" licenses apply to one of the 20 territorial telecom circles (which roughly correspond to the boundaries of a state) or the large metropolises.
- Category "C" licenses are for individual cities.

The policy removed restrictions on the number of licenses a single company could acquire or the number of licenses that could be held by different companies in the same area. Licenses would be issued for 15 years and ISPs would be free to set their own tariffs. In addition, ISPs could lease transmission network capacity from private basic service operators, the railways, the state electricity boards and the National Power Grid, thus ending the DoT's monopoly on domestic long distance data networks. ISPs could also build their own transmission networks, upon approval from the DoT.

When the *New Internet Policy* went into effect on 6 November 1998, companies rushed to acquire licenses to provide Internet services. As shown in Figure 14.2, the growth in the number of licenses issued remained linear through the end of 2000, when the total exceeded 400. New license issuance came to a standstill in 2001, as India felt the effects of the bursting of the global dot-com bubble and the cooling incentives and declining support for new entrants into the Internet services market. The decline in 2002 reflects the issuance of an exit policy for license holders that would enable them to surrender their licenses by paying a surrender charge of 5% of the Performance Bank Guarantee (PBG) (Anonymous, 2002).

The actual number of operating ISPs at any given point in time is difficult to determine, although an estimate is given in Figure 14.2. The gap between the number of companies licensed to provide Internet services and the number that are operational can be explained in a number of ways. Part of the explanation is that offering Internet services is considerably more difficult than acquiring a license to do so. Many ISP aspirants underestimated the effort involve in establishing a viable business. But other factors played a role.

One factor appears to be the role played by the state-owned ISPs. As the only commercial ISP from 1995 to 1998, VSNL was able to accumulate experience, establish an extensive infrastructure for ISP services and gain early and elite users as customers. In addition, since VSNL was the monopoly provider of international Internet leased lines, it was able to hinder the growth of ISPs by delaying their applications for leased lines and by offering them connectivity at rates that squeeze their profit margins. VSNL's internal costs for connectivity are lower than what it charges ISPs. Consequently, if VSNL and the ISPs charged comparable rates to the end user, VSNL would have a wider profit margin and could thus tolerate lower prices than could the ISPs (Ganapati, 1999). For example, in August 1999, VSNL received permission to reduce the rate it charged corporate customers for international Internet connections by 15%, while leaving the rates for private ISPs unchanged (Mani, 1999; Mohan, 1999). These lower rates would be made available to recognized educational institutions, government organizations, newspapers and news agencies and corporations located in a business cluster, such as a software technology park.

Figure 14.2
ISP License Holders

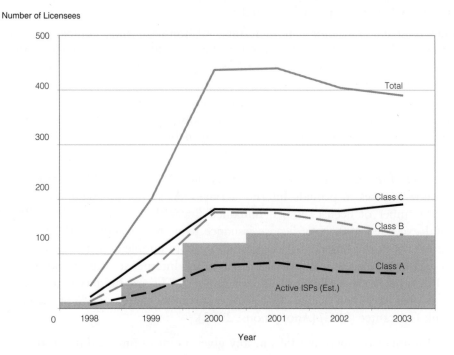

Source : Anonymous, 1999a, b, c, 2001a; Antonelli, 1986

A similar situation existed in Delhi and Mumbai, where MTNL was a monopoly provider of phone services. In July 1999, MTNL reduced its prices for certain institutional end users by 15% (Anonymous, 1999d), while the profit margins of competing ISPs in these cities were less than this amount. Consequently, MTNL was either subsidizing a service it was offering below cost, or its costs were lower than the ISPs. Either answer could be viewed as an anti-competitive practice.

On 3 January 2001, VSNL introduced a popular pricing package called "Monsoon" which reduced dial-up costs by 50% (Sengupta, 2001). This reduction came at a time when many ISPs were considering increasing access rates to improve their financial footing. The price reduction and resulting price-war among ISPs forced many to curtail their plans for expansion, or cease service altogether in certain cities (Goswami, 2001; Raman, 2001). In contrast, VSNL experienced steady increases in its subscriber base and acquired a reputation for high quality connections (Anonymous, 2001b).

The intensive competition has continued to the present. According to reports in August, 2003, VSNL, MTNL and BSNL were the first, third and fourth largest ISPs in India, together having 45% of the nation's subscriber base. With the exception of Satyam Infoway (Sify), the second largest ISP with 17% of the subscriber base, the private ISPs lag the public ones by a considerable margin. Industry observers attribute this to a combination of network capacity, accessibility and aggressive marketing (Kumar, 2003). As Figure 14.2 shows, the competition does not appear to have led to a substantial reduction of the total number of active ISPs. It is quite possible, however, that the slight decline in 2003 will intensify in 2004 and 2005, leading to a competitive, but less diverse market for Internet services in the years to come. What, if anything, policy-makers and regulators should do to address this scenario is likely to be a subject of discussion.

Figure 14.3
Estimated Number of Internet Users in India

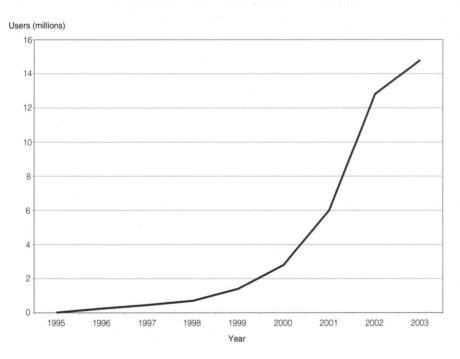

One of the consequences of the aggressive competition has been dramatic growth of the Internet as measured by number of users, number of ISP points of presence and number of cities served. Figure 14.3 shows the growth of the estimated number of Internet users in India. Figure 14.4 shows the growth of the number of ISP points of presence.

From 1998–2003, the number of cities with points of presence grew by an order of magnitude, from approximately 45 to nearly 500, as shown in Figure 14.5 also reveals the growing competitiveness in the Indian market for Internet services. Markets with more than 10 competing ISPs emerged for the first time in 1999 (Hyderabad). In 2000, five cities were serviced by 20 or more ISPs—Bangalore, Chennai, Delhi, Hyderabad and Mumbai, with the greatest concentration in Delhi and Mumbai. A year later, Kolkata, Ahmedabad and Pune were added to this list.

A comparison of Figures 14.3–5 illustrates other significant points about Indian Internet expansion. The rate of increase of total cities served in 2001–02 is greater than the rate of increase of ISPs, which in turn is greater than the rate of increase of users. The underlying reason is that ISPs tended either to offer localized service (Class "C" or "B" licenses), or to target primarily the larger metropolitan areas. The exception was BSNL. While 2001 was a year in which private ISPs did little to expand service into previously un-served areas, most of the growth in 2002 occurred through BSNL's efforts to expand the number of cities served. During 2001 and 2002, multi-year efforts by BSNL to implement a national Internet backbone to provide Internet services in all districts finally began to bear fruit. Most of the expansion in points of presence (Figure 14.4) and in cities (Figure 14.5) is due to the rollout of the Sanchar Net Internet service in districts nationwide (Anonymous, 2003b). These points of presence do not generate the numbers of users that points of presence in the more populous cities do, a fact reflected in the more modest rate of growth of users (Figure 14.3).

Figure 14.4
Estimated Number of ISP Points of Presence

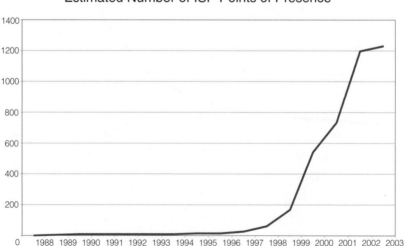

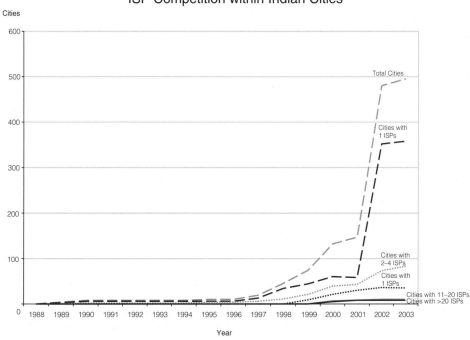

Figure 14.5
ISP Competition within Indian Cities

With points of presence in over 400 cities, BSNL's Sanchar Net provides service to eight times the number of cities of the next most widespread ISP (Sify). This fact reflects a cornerstone of policy-makers' attitudes towards telecommunications services in general. While policy-makers have been eager to introduce market forces into the telecommunications industry, the dedication to achieving universal service remains a prominent part of policies. While market forces have succeeded in driving Internet expansion in more heavily populated areas, a universal service objective and government investment have been necessary to push the technology beyond major cities. The universal service objective has been behind the DoTs expansion of telecommunications infrastructure from a teledensity of 1.5 lines per 100 people in 1998 to 6.2 in 2003 (Anonymous, 2003a). It has also motivated DoT/BSNL efforts to build the national Internet backbone (Agarwal, 1999; Rajawat, 2001) and Sanchar Sagar's (Anonymous, 2003c; Kumar, Neeraja, 2000) two major efforts to build national Internet infrastructure and roll out service to all district headquarters.

5. Discussion

A framework for analyzing strategies of ISPs has been suggested by Wierstra and others, based on their analysis of ISPs in Europe (Wierstra *et al.*, 2001). They identify three principal types of providers of Internet service. *Incumbent telecommunications companies* are characterized by their ownership of national or international infrastructure, a highly vertically integrated set of services and a broad portfolio of basic and value-added services which they offer to a broad spectrum of residential and business customers. *New carrier companies* are characterized by high capital investments to build new infrastructure which is used primarily for providing IP-transport and intranet services to businesses, including large national firms (LNF), multinational enterprises (MNE) and ISPs. *Small, regional ISPs* do not generally own their own infrastructure, are not vertically integrated and offer a portfolio of Internet access services and consultancy to residential and small and medium sized enterprises (SME).

These companies can compete on any combination of five strategy dimensions. *Factor inputs* relate to the technology employed and to specific capital and labour inputs. The *functional value* chain refers to how vertically integrated a company is. The *product portfolio* reflects the breadth and depth of services offered, with a basic portfolio consisting of basic IP transport, web hosting and e-mail. *Pricing policies* include both prices in absolute terms and whether the pricing policy is a flat fee or usage based. *Market segmentation* includes the geographic scope (regional, nation-wide, or international) and residential vs. business user categories.

Based on these factors, the authors suggest the following competitive balance between the incumbent telecommunications companies and the small, regional ISPs.

- These companies compete primarily in the residential user market rather than the large national firms and MNEs.
- Incumbent telecommunications companies may enjoy competitive advantages through:
  o a nation-wide supply of POPs,
  o established relationships through supply of telephony,
  o cost advantages due to ownership of infrastructure,
  o economies of scale in development of content as a result of broader user bases.
- Small ISPs may enjoy competitive advantages through:
  o Customized support services.
- A weakness of the incumbent ISPs is that they use relatively old technology and may face threats from new entrants in the market for backbone capacity.
- The regional ISPs may face weaknesses from their dependency on the incumbent telecommunications for infrastructure and their inability to gain economies of scale.

To some extent, the evolution of the Internet services market in India supports this characterization of the competitive landscape. The competition between the state-owned and the private ISPs for residential customers is intense. The incumbent state-owned providers (VSNL, MTNL and BSNL) have exhibited an aggressive strategy of reduced costs and bundling of services that leverages their control of existing infrastructure and existing relationships with customers. According to the Internet Service Providers Association of India (ISPAI), the incumbents bundle their Internet and telecom services and force unequal interconnect terms on private players (Nair, 2003). It is not clear, however, that nation-wide coverage is a significant factor, particularly for MTNL which is limited to two major metropolitan areas. The fraction of users who depend on the ability to access their accounts from numerous cities within India is likely to be small.

A substantial portion of the competitiveness of the state-owned providers has been their aggressive build-out and upgrading of infrastructure, thus broadening the geographic scope of coverage, improving the quality of Internet service, and reducing the unit costs of network capacity. In this manner, the state-owned ISPs have worked vigorously to mitigate a potential weakness identified in the Wierstra analysis framework.

While it partially helps explain the evolution of Internet services in India, Wierstra *et al.*'s (2001) framework is incomplete in several respects. First, the Internet service providers of India are not so easily placed into the "small, regional ISP" and "nationwide incumbent telecommunications company" categories. First, of the incumbent telecommunications companies, only VSNL and BSNL offer nationwide services. In addition, a number of private ISPs, such as DishnetDSL and Sify, hold Class A licenses and some have begun to build their own infrastructure and enter the telephony markets. These companies have begun to exhibit more of the characteristics of the incumbents and enjoy some of their strategic benefits.

Second, there are factors besides those listed above that have strongly shaped the competitive environment. The authors acknowledge that a company's competitive position is affected by other market conditions, such as demand technological factors, policy and regulation, and the number and strategies of competitors, though such factors are not explicitly a part of their framework. Yet it is these that have had such a strong impact on the Indian ISP landscape. In particular, the Indian government's decisions to gradually replace monopoly rights with competitive environments has invigorated these environments yet left the incumbent providers with sufficient resources to compete aggressively with new entrants. The competitive environment has driven the state-owned providers to reduce their costs, increase reliability, and offer an innovative set of services (Kumar, 2003).

Finally, the expansion of BSNL to over 400 cities throughout the nation cannot be explained solely by a desire to increase the number of subscribers. Most ISPs have concentrated their efforts in the densely populated areas where per-subscriber costs are lower. BSNL's strategy reflects government policy, particularly that of universal service, and the need to bring the Internet to all parts of the country if India is to become a "major IT power".

6. Conclusion

Though India's experience with the Internet and the telecommunications sector shares many similarities with the experiences of other countries, its unique history, size, and demographics give it distinctive qualities which offer insights into Internet diffusion. While the discussion here is limited to government policy and selected aspects of the expansion of Internet services, it underscores the critical nature of one of the factors related to Internet growth—the relationship between policy-makers, state-owned telecommunications providers, and private Internet service providers.

India's experience offers a number of insights into and lessons for the development of the Internet. This country's experience has been, on the whole, successful, and offers one model of how countries can transition from Internet laggard to one with a vibrant Internet presence.

First, government policy makes an enormous difference in Cyberspace. In spite of those who claim that the Internet transcends national boundaries, the Indian experience provides an excellent illustration of how fundamental, focused changes in policy and legislation can unleash forces that can accelerate Internet diffusion. A number of factors shaped telecommunications policy, but the legislative acts and regulatory decisions of the Indian Government have been quite coherent and forward looking. Global trends, Indian leadership, and domestic conditions converged to support changes in policy that have benefited the Internet greatly. Among the key changes were the creation of a regulatory regime that grants start-up companies rather easy entry into the Internet services market, and reasonably fair access to existing infrastructure.

Second, the nature of the relationship between the government, the state-owned telecommunications service provider(s), and the private sector is a critical variable. While many variations on this three-way relationship exist, India offers one example in which the relationships have a tension that ultimately helped the Internet grow. Indian policy-makers have helped shape an environment in which the incumbent telecommunications providers have neither withered away in the face of competition by start-ups nor overpowered the latter. The result has been a strong market ultimately benefiting the end users.

Third, Indian policymakers have been rather successful using both market and non-market forces to encourage the growth of the Internet. Gradual but deep changes in the regulatory environment have created a highly dynamic milieu for Internet services. At the same time, major infrastructure projects and mandates for publicly-owned providers have enabled the expansion ISP services to parts of the country not well served by the private ISPs. An important lesson is that both public and private initiatives play important roles; neither should be relied on exclusively.

Barriers to the Internet remain significant. Poverty remains extensive. While teledensity figures have improved considerably, they remain well below those of developed countries. The climate for foreign direct investment is questionable. The government has not always been able to carry out large-scale infrastructure projects in an efficient, timely manner. Whether the creative tension between Internet service providers can be maintained is an open question. With only approximately 1.5% of the population using the Internet, India still has a long way to go to become the "IT Power" envisioned by Vajpayee. However, India has made enormous strides. Current trends and policies are promising.

References

Agarwal, P. K. (1999) Building India's National Internet Backbone, *Communications of the ACM*, **42**, 6, 53–58.

Anonymous (1997) Guidelines and General Information For Internet Service Provider (ISP) (No.845–51/97-VAS), Department of Telecommunications.

Anonymous (1998a) ERNET India, Department of Information Technology, www.eis.ernet.in/about.html.

Anonymous (1998b) India Internet Users Top 100,000, Seen Surging, Reuters, Apr. 17.

Anonymous (1998c) Information Technology Action Plan. National Task Force on Information Technology and Software Development, New Delhi, July 4, it-taskforce.nic.in/infplan.htm.

Anonymous (1999a) DoT Issues 132 ISP Licences, *The Hindu Business Line*, Aug. 1.

Anonymous (1999b) Internet in India: A Sunrise Industry, *India Internet World* 1999, www.iworldindia.com/netinindia/netinindia.html.

Anonymous (1999c) ISP Licences Issued, Department of Telecommunications, www.dotindia.com/.

Anonymous (1999d) MTNL Cuts Net Tariffs; VSNL, Dishnet Join Hands, Computers Today, July 16-31, www.india-today.com/ctoday/16071999/buzz3.html.

Anonymous (1999e) New Telecom Policy 1999, Government of India, www.trai.gov.in/npt1999.htm.

Anonymous (2000) The Information Technology Act, 2000, Ministry of Law Justice and Company Affairs, The Gazette of India 33004/2000 Part II, 2–32.

Anonymous (2001a) List of ISP Licences Issued (As On 30.9.2001), Department of Telecommunications, www.dotindia.com/isp/ispindex.htm.

Anonymous (2001b) VSNL Revises Net Tariffs; Unlimited Access Goes, *The Hindustan Times* (New Delhi), Sep. 7.

Anonymous (2002) Exit Policy for ISPs, Department of Telecommunications, www.dotindia.com/isp/exitpolicy.htm.

Anonymous (2003a) Network Status (October 2003), Department of Telecommunications, www.dotindia.com/networkstatus.htm.

Anonymous (2003b) Sanchar Net, Bharat Sanchar Nigam Ltd., www.sancharnet.in/.

Anonymous (2003c) Sanchar Sagar Project, Bharat Sanchar Nigam Ltd., www.bsnl.co.in/company/sanchar.htm.

Antonelli, C. (1986) "The International Diffusion of New Information Technologies," *Research Policy*, 15 139–147.

Dossani, R. and Manikutty, S. (2002) An Institutional View, in: Dossani, R. (ed.) *Telecommunications Reform in India*. Westport: Greenwood Publishing Group.

Ganapati, P. (1999) Where Have All the Cowboys Gone?, Rediff on the Net, Aug. 21, www.rediff.com/computer/1999/aug/21isp.htm.

Goswami, R.R. (2001) ISPs Find Going Tough in Gujarat, Shut Shop, *The Economic Times* (New Delhi), Sep. 28.

Kumar, A. (1998) Presentation, India Internet World'98, New Delhi, Aug. 25.

Kumar, N. (2000) Sanchar Sagar Phase II to be Completed by March 2001, *Financial Express* (Bombay), Apr. 21, www.financialexpress.com/fe/daily/20000421/fco21022.html.

Kumar, P.V. (2003) Empire Strikes Back: VSNL Beats Private ISPs in Numbers Game, *Financial Express* Internet edition, Aug. 20, www.financialexpress.com/fe_full_story.php?content_id=40397.

Mani, A. (1999) TRAI Clear VSNL Internet Tariff Cut Despite Opposition, *The Hindu Business Line*.

Mohan, R. (1999) Cheaper VSNL Net Rates for Corporate Category, *The Hindu Business Line*, Aug. 6.

Nair, V. V. (2003) BSNL, MTNL Monopoly in Net Services Alleged: Private ISPs Seek TRAI Intervention, *The Hindu Business Line* Internet edition, June 24, www.blonnet.com/2003/06/24/stories/2003062402410100.htm.

Pai, U. (1998) India Prepares For Flood of ISPs, *TechWeb*, Aug. 18, www.techweb.com/wire/story/TWB19980818S0002.

Rajawat, K. (2001) BSNL Completes Phase I of National Internet Backbone, *The Economic Times of India*, July 18.

Raman, K. (2001) VSNL Pricing Puts ISPs on Backfoot, *The Hindustan Business Line*, June 16.

Rambabu, G. (2004) On the Right Track, *The Hindu Business Line*, www.thehindubusinessline.com/2004/01/09/stories/2004010900230400.htm.

Sengupta, S. (2001) VSNL Slashes Dial-up Rates by 50%, *The Economic Times* (New Delhi), Jan. 3.

Vajpayee, A.B. (1998) IT is India's Tomorrow, Financial Times India: Information Technology. Supplement to the *FT-IT Review*, Dec. 2, 1998, 5.

Wierstra, E., Kulenkampff, G. and Schaffers, H. (2001) A Framework for Analyzing Strategies of Internet Service Providers, *Netnomics: Economics Research and Electronic Networking*, 3, 1, 35–65.

Wolcott, P. and Goodman, S. E. (2003) India: Is the Elephant Learning to Dance?, *Communications of the AIS*, 11, 32, 560–646, cais.isworld.org/articles/default.asp?vol=11&art=32.

Wolcott, P., Press, L., McHenry, W., Goodman, S. E. and Foster, W. (2001) A Framework for Assessing the Global Diffusion of the Internet, *Journal of the AIS*, 2, 6, jais.isworld.org/articles/default.asp?vol=2&art=6.

About the Author

Peter Wolcott is an Associate Professor of Management Information Systems at the University of Nebraska at Omaha. He has long-standing interests in the international dimensions of information technologies. His current research projects examine municipal e-government, and the global diffusion of the Internet and electronic commerce. Other research interests have included high-performance computing export control, high-performance computing of the former Soviet Union, and data warehousing.

His work has been published in *Communications of the ACM*, *IEEE Computer*, *Journal of the AIS*, *Communications of the AIS*, *The Information Society*, *The Journal of Data Warehousing*, and other journals and proceedings.

He earned his Ph.D. at the University of Arizona in Business Administration (Management Information Systems) in 1993.

15. Information Technology for Development in India— The Kerala Experience

K. G. K. Nair

Department of Business Administration
College of Engineering, Thiruvananthapuram-695016
Kerala, India
gopalmanjusha@yahoo.com

P. N. Prasad

State Institute of Rural Development
E.T.C. (P.O), Kottarakara-691531
Kerala, India
pnprasad_2000@yahoo.co.uk

Abstract

Kerala, one of the states of India, is a region known for its high levels of achievements in education, healthcare and social justice despite low per capita income. However, the promotion of Information Communication Technologies (ICT) was delayed in the state due to the widespread misconception in the minds of the people that its increased application may replace the work force, resulting in loss of employment. It was only in 1998 that the government of Kerala formulated its first Information Technology (IT) policy. The IT policy was revised in 2001, incorporating many new features for accelerating the promotion of IT industry in the state. The article highlights the salient features of the IT policies in the state. The government's initiatives, achievements and limitations in the implementation of IT policy are also discussed in detail in this article. The state has achieved reasonable success in e-governance with several model projects having been initiated for the benefit of the common man. However, in spite of several intrinsic advantages, the development of IT industry in the state has not been commensurate with its potential. The paper describes case studies of a number of innovative IT projects in the state that hold substantial potential for bridging the gap in the digital divide. The findings of the study also reveal the need for a suitable strategy for the promotion of IT industry.

1. Background Information about Kerala

Kerala is one of the states of India, with an area of 38,863 sq. km and a population of 31.8 million, of whom nearly 75% live in rural areas. Apart from the national government, the state has an elected state legislature and a three tier structure of elected local bodies for the governance. There are 991 grama panchayats (village level administrative set-ups) for rural areas and 53 municipal councils and 5 corporations for urban areas. Kerala, known as "God's own Country" for its natural beauty, is one of the world's top tourist destinations; having been listed as one among the world's 50 "must see places" of a life time (National Geographic Traveller, 1999).

The state takes a proud position in the indices of human development measured in terms of life expectancy, education and healthcare. The literacy in the state is 90.92%, male literacy being 94.20% and female literacy being 87.86%. According to the 2001 census, the state occupies the foremost position in the country in education, healthcare and population control (Government of Kerala, 2001a). The achievements in these sectors are even comparable to some of the developed countries in the world.

2. Introduction

In the context of developing countries, IT is seen as one of the most significant forces of modernization. In the global "information society" the various indicators of diffusion of IT are a characteristic of development (Avgerou, 2000). However, for more than a decade the IT revolution in developed countries was viewed with suspicion by both the trade unions and the people of Kerala, because they considered it relevant only to rich countries that have a shortage of manpower. The trade unions were against any positive action to encourage and promote this technology. They feared that computerization would ultimately lead to the destruction of job opportunities, which will be fatal for a state like Kerala, where unemployment is the main problem. They believed that the benefits of IT would remain confined to the higher classes of the society. Consequently, the government shied away from the promotion of IT, even when intensive promotional activities were going on in neighbouring states like Karnataka, Andhra Pradesh and Tamil Nadu.

It took some time for the state to realize that IT is relevant to all countries, irrespective of their level of economic development or varied local problems. Before 1995, the Government of Kerala did not make any serious attempt to promote IT, probably due to the widespread misconception in the minds of the trade unions and common people that the increased application of IT may replace the work force, resulting in loss of employment. However, the growth of IT in nearby states and all over the world, along with the lucrative jobs obtained by Keralites in IT outside the state, were eye-openers to the disbelievers. The potential of IT in creating large-scale employment was realized by the state, albeit a little late. Though the government had initiated some activities in the early 1990s, it was only in 1998 that the government formulated a comprehensive IT policy. The IT policy was revised in the year 2001, incorporating many new features, with a view to increase the employment opportunities through the application of IT.

3. IT Policy 1998

The Kerala state government realized the enormous potential of IT, not only as a tool for improving governance and creating more jobs, but also more significantly as a means to enhance the standard of living of the people. The objective of the IT policy was to increase the application of IT in all walks of life, enhancing the IT industry base, creating a robust state information infrastructure and creating human resources for IT development (Government of Kerala, 1998a). The IT Policy mission was spelt out as follows:

- Upgrading of the standard of living of the people of the state through use of IT in all sectors as a tool to enhance productivity, efficiency and optimal utilization of resources, and through full exploitation of the employment potential of the IT sector;
- Establishment of a State Information Infrastructure (SII) comprising a high speed broadband communication backbone, nodes, access network, distributed data warehouses and service locations to cater to the needs of trade, commerce, industry and tourism and also to enhance the delivery of government services to the people;
- Establishment of the state as a fertile location for the growth of the IT industry, facilitating the flow of investments from within the country and abroad, achieving in the process the rapid growth in domestic and export earnings;
- Development of human resources for IT through increased use of IT in educational institutions and through academic and training programmes that improve the employability of educated youths in the IT sector;
- Facilitation of decentralized administration and empowerment of people through the application of IT.
- Components identified for the speedy and effective realization of the policy were:
- Diffusion and dissemination of IT
- Enhancement of the industry base of IT
- Creating a robust information infrastructure for the state
- Human resource development for IT

In line with the 1998 IT policy, a high power committee was constituted under the chairmanship of the chief minister of the state to co-ordinate and direct strategies to achieve rapid penetration and effective use of IT in all sectors. Three broad areas have been identified and separate task forces were constituted for:

- Human Resources Development and IT Dissemination
- Enhancing the IT Industry base
- IT implementation in government

A state advisory council on IT was also constituted with N. Vittal (Former Secretary, Department of Electronics, Government of India) as Chairman. The council, in its report submitted to the government, suggested that the IT policy objectives must be achieved within 1,000 days starting from 1 December 1998 (Government of Kerala, 1998b). In tune with this policy, the government has initiated many programmes and projects for the development of the IT base, especially for e-governance in Kerala.

4. IT Policy 2001

Consequent upon the general election that took place in Kerala, a new government was formed in the state in 2001. The new government revised the IT policy in December 2001. The new policy did not envisage any drastic change in the attitude of the government. However, the focus has been changed from e-governance to the development of the IT industry (Government of Kerala, 2001b). The policy initiatives outlined in the new IT policy document comprised a three pronged strategy aimed at:

- Creating an appropriate pro-business, pro-enterprise, legal, regulatory and commercial framework to facilitate the rapid growth of the IT industry in the state;
- Establishing Kerala as a global centre for excellence in human resources, through the creation of a large pool of diverse, multi-skilled, technically competent manpower in the state; and
- Establishing an internationally competitive business infrastructure and environment for the IT industry in the state, on par with the best facilities and practices world-wide.

4.1 Features of the IT Industry Policy

Kerala's IT Industry Policy—2001 incorporates the first tailor-made regulatory framework for the IT industry, wherein software companies would be deemed as establishments under the "Kerala Shops and Commercial Establishments Act, 1960". With specific exemptions provided under this act, Kerala shall be the first state to permit three-shift operation, permission for women working in the night shift, and flexitime working. Further, the IT industry has been deemed a Public Service Utility under the Industrial Disputes Act, 1943. The IT policy also permits self-certification in respect of nine statutory Acts. It also recognizes the practice of providing for part-time working and teleworking, both of which additionally enable IT Enabled Services (ITES) operations.

The Kerala Government's IT Industry Incentive Scheme 2001–03 encompasses a unique, comprehensive package of incentives and policies for IT products, software and the ITES industry. The IT Industry Employment Promotion Scheme offers an incentive scheme in which it is simple to participate that is truly a win-win strategy for government and industry, wherein incentives are structured in favour of higher employment providers.

Other incentives include a unique "Early Bird" incentive that offers a one time incentive payment of Rs. 50 lakhs[45] to the companies that set up operation in Kerala before 30 June 2002, and provides employment for 250 persons. Also, for the first time, a state Government shall directly provide incentives for the acquisition of quality certification CMM level 5 (Capability Maturity Model), COPC-2000 (standard of the Customer Operations Performance Centre), eSCM level 3 (capability of the Services Capability Model) of the companies within the state.

The IT policy aims to generate a trained pool of manpower, with the skills and aptitudes appropriate to the ITES industry requirements in particular, and the IT industry as a whole. A comprehensive and focussed state-wide HRD programme is planned to achieve this objective. Bridge programmes for engineering graduates, communicative English, soft skills, accent neutralization and ITES sub-domain level training shall be the initial thrust areas. These programmes shall be integrated into the mainstream collegiate education system, including continuing education programmes, in the state. From a long term perspective, spoken and written English shall be emphasized at the school level.

5. Policy Initiatives—Case Studies

5.1 IT Implementation in Government Departments

The task force on IT implementation in government identified the departments in which there is a high level of interaction with the public, where computerization shall facilitate increased revenue collection (Kerala Government, 1998c). The task force identified and listed 31 such departments for IT implementation in the first phase. The government has earmarked 3% of the state budget for implementation of the IT action plan. The treasury department, the registration department, the civil supplies department and the motor vehicles department have implemented model projects in selected locations. By computerizing the functions like office procedures, documentation and delivery of services, the common people are getting the services more efficiently. The remaining departments have also procured necessary hardware and software and the officials are being trained in computer applications.

The case study on computerization of government departments shows that for many people as well as institutions, computerization means procurement of hardware only. In the majority of cases, people do not go beyond using standard application packages, resulting in gross under-utilization of computers. Computers are being used

---

[45] A lakh is an Indian numerical unit equivalent to 100,000. Hence 50 lakhs is 5 million.

as a personal tool of a few officers rather than as a tool to improve system performance. In some departments, computers are not being made available to the trained personnel. Time and resources are seldom spent to develop application packages for specific purposes, which are essential for any meaningful computerization programme.

There is considerable enthusiasm at the department level, driven by forward-looking individuals, but a cohesive movement is lacking. For a meaningful computerization of government departments, it is required that the computers should be maintained properly by the trained manpower. Also, software packages suitable for specific office applications should be designed for the delivery of services to the citizens. Shortages of consumables, service problems, lack of trained manpower, etc. should not come in the way of the proper functioning of the computers.

5.2 Information Kerala Mission

The IT policy-1998 of the Kerala state government focused on the application of IT in a decentralized democracy. Accordingly, computerization of all grama panchayats was given a high priority in the IT policy. The Information Kerala Mission, which was set up in 1998, is a trend-setter in the deployment of technology at the grassroots level and a model for participatory governance through the effective use of IT. The project aims at modernization and integration of government functions using IT by networking and computerising the local self-governments to expedite project implementation and transactions like issue of certificates, licensees, tax collection, etc.

The Information Kerala Mission has developed the software for the effective functioning of grama panchayats, which includes software for monitoring of plan implementation and certificate issuing. The project also envisages massive awareness programmes through training and orientation at all levels of village, district and state government bodies. The government has achieved significant success by developing Kumarakom panchayat in Kottayam district and Vellanad panchayat in Thiruvananthapuram district as role models. For the first time in the country, a project for establishment of computer information network covering all the 152 development blocks and district headquarters has been completed. Thus, the computerization of local self-government has been progressing at a fast pace. The various services from these offices like issue of certificates, licenses, tax collection, etc. are expected to become a reality in the near future and the rural population is also being brought under the ambit of application of IT.

5.3 FRIENDS

FRIENDS (Fast, Reliable, Instant, Efficient, Network for Disbursement of Services) has been set up in every district head quarters by the IT department. FRIENDS is a unique project taken up by the government to provide the benefits of IT to the common man. This offers the citizen an integrated point for utility payments for specific services such as electricity charges, water charges, property tax, professional tax, building tax, etc. The server of the FRIENDS centre is being linked to the server of the concerned departments for updating on a real time basis. A friendly "Help Desk" helps the customers to fill up forms and clarify their doubts. The project when completed is expected to benefit 12 million people of the state i.e., about 40% of the population.

5.4 SEVANA

With a view to disseminate increased application of IT to the rural people, in association with the library council, the IT department has started a novel project called SEVANA. The project envisages converting about 400 libraries in rural areas into IT dissemination centres by providing free Internet connections so that they can function as rural information centres. A software package named SEVANA provides information on various government schemes, programmes, general information on local bodies, links to important sites, and other important facts relevant to the rural population. The pilot project, which was implemented in Kallara Panchayat in Thiruvananthapuram district, has been functioning very well. The centre became a point of convergence for villagers seeking information services and an entry into the world of computers and computer literacy. The people from rural areas are able to get various utility services through the Internet. Training for Internet awareness is also being organized in these centres.

5.5 PEARL

PEARL (Package for Effective Administration of Registration Laws) is a joint project of the Registration Department, IT Department and National Informatics Centre. The project is being implemented as a major initiative of the government towards providing the benefits of IT to the common man. The National Informatics Centre is the overall consultant for the project, including its design and development. PEARL aims at providing a transparent, efficient and vibrant public interface thereby bringing in efficiency and motivation at the Sub Registry office level. This dispenses with the dreary manual routine of filing, searching, accounting, reporting etc. This in turn should lead to a drastic reduction in the number of opportunities for corruption, thereby enhancing the satisfaction and awareness at the general public level. This also provides the common man with the power of getting prompt and reliable response from the department.

The project for computerization of the Registration Department of the state aims to reduce document registration time from two weeks to a few hours and bring more transparency into the functioning of the state-wide department. The main feature of the project is that the existing registration system is being replaced by a system of online processing. For this, every sub registry office would have a local area network with adequate terminals for customer servicing. The computerization will cover registration of all categories of documents, issue of encumbrance certificate and issue of certified copies. The advantages of PEARL include registration of documents within one hour as against two weeks, and issuing of encumbrance certificates in 10 minutes instead of the existing situation, where it takes up to one week.

Certified copies of documents can be issued within a minimal time, as maintenance and filing of documents is being done electronically. Data warehousing and retrieval system are an integral part of the package. The entire process of document writing and registration is substantially simplified, as document writers are no longer required to prepare filing sheets, in the new computerized system. Moreover, the implementation of the project would pave the way for transparency in the registration department and enable more effective monitoring. It would also enhance the revenue collection in the state.

5.6 Indian Institute of Information Technology and Management—Kerala (IIITM-K)

IIITM-K is conceived as a centre of excellence, on a par with the IITs (Indian Institutes of Technology) and aims to train skilled professionals in leading IT and Management disciplines. The institute is located in the campus of Technopark—India's most advanced destination for IT. This institute, the establishment of which has been envisaged in the IT Policy of the Government of Kerala, is conceived as a national level institution and will offer a range of Masters Programmes in leading IT disciplines. An important feature of the institute is the significant levels of industry participation in the governance of the institute and in the design of the curriculum. The institute will offer two sets of academic programmes—Masters Programmes and Short Term Professional Programmes. The full time Masters Programme will be the core programme of the institute. This programme—the Masters Programme in Information Technology (MITech)—will be structured so as to sensitize the participants to key issues of IT—Business and Management. In addition to these academic programmes, the institute will also offer professional testing and accreditation services for private training institutes.

5.7 Rights of Way Policy—Drive for IT Infrastructure

The government of Kerala declared the rights of way policy, permitting any private or public sector infrastructure providers who propose to lay fibre optic cables to have right of way over public properties including national and state highways, village roads and other public properties on a non exclusive basis. Accordingly, the government has already finalized agreements with infrastructure providers and the work has been progressing. This should enable the construction of a fibre optic backbone right across the state and would ensure that high speed Internet connectivity is available in all district headquarters.

5.8 Information Kiosks

To own a PC, a modem and a telephone connection, the user end requirements of Internet, is a distant dream for the common man in Kerala. However, it must be remembered that in Kerala, the common man enjoys the communication facility not by owning a telephone, but through the telephone booths available in every nook and corner of the state.

One of the targets of the IT policy was the setting up of Internet kiosks in every Panchayat ward, accessible to any member of the public. The Kerala State Electronics Development Corporation (KELTRON) has already set up Keltron Information Kiosks (KIK) in three locations in Trivandrum district and it proposes to set up similar kiosks in all the other districts of the state. The objective of the centre is to enable the government to provide quick service to the common man. The focus of KIK is to generate content that is locally specific. KIKs are value added cyber cafés providing various services related to government through the Internet and Local Area Network (LAN) that caters to the everyday needs of the masses. The KIK envisages an E-governance grid by networking government departments, institutions and other agencies involved in governance. The goal of the E-governance grid is on the one hand to enable the government to improve delivery of various services and on the other to empower the citizens to actively participate in the digital revolution, both as clients of government services and also partners in creating, maintaining and disseminating information, which is locally relevant and specific (Kumar, 2001).

5.9 Akshaya Computer Project

Akshaya computer project, aimed at bridging the digital divide was inaugurated in the state by the President of India, Dr. Abdul Kalam on 18 November 2002. The highlights of the project are:

- At least one person in every family will be made IT literate
- Multipurpose communication technology centres will be developed within 2 Km radius of any household through private sector initiatives
- Focused campaign on the benefits of IT to the masses supported by skill development.

5.10 Development of State Information Infrastructure

The State Information Infrastructure (SII) shall endeavour to provide robust and seamless connectivity to industry across the state. At present, Kerala enjoys the highest telephone density in the country. 100% of the 982 telephone exchanges in the state are digital, covering all 14 district headquarters, 63 taluk headquarters, and 1468 panchayats/villages. The Bharat Sanchar Nigam Limited (BSNL) telephone exchange is located in panchayat headquarters in a majority of panchayats and the balance are within no more than three kilometres of the nearest telephone exchange. 98% of the exchanges are networked through optic fibre cable (OFC). 266 exchanges currently support ISDN service.

Videsh Sanchar Nigam Limited (VSNL) has made substantial investments in its Kochi gateway where two submarine cables-SEA-ME-WE (connecting Australia, South East Asia, the Middle East and Western Europe) and SAFE (connecting South Africa to the Far East) land. Kochi is one among two locations nation-wide, which has submarine cable landing. With 10 GBPS bandwidth supported, it currently handles about 70% of the nation's data communication traffic. Sourcing bandwidth from

VSNL at Kochi and Thiruvananthapuram provides significant savings *vis-à-vis* most other locations in the country.

It is noteworthy that the government has achieved good progress in achieving the projected target of state information infrastructure development. The completion of ongoing projects in IT infrastructure will provide connectivity at 2 mbps to any user in the major cities of the state and in multiples of 64 kbps in other parts of the state. The submarine cables SEA-ME-WE and SAFE are being extended from Cochin to other parts of Kerala under project "100 percent digital Kerala" to provide band-width to any user as per his requirement. This will make the state a highly attractive base for the high-end IT companies.

5.11 Promotion of IT Industry

The government of Kerala lists the following advantages for the state as an IT desti-nation (Government of Kerala, 2002).

5.11.1 The Technology Advantage

- 100% of 988 telephone exchanges are digital
- 98% of telephone exchanges connected by OFC to the National Internet Back-bone (NIB)
- Private companies like Reliance, Bharti and Asianet are laying their own OFC backbones
- Highest telephone density—7 per 100—India's 2005 target
- "SEA-ME-WE-3" and "SAFE" submarine cable landings (1 of the 2 states in In-dia to have two submarine cable landings)
- 15 GBPs bandwidth supported
- VSNL's primary international gateway in India is in Kochi, Kerala

5.11.2 The Human Resources Advantage

- Highest density of science and technology personnel in India
- Lowest employee attrition rate in the country—less than 5%
- 72 engineering colleges in the state
- Database of readily employable graduates enabling interested ITES companies to access the best of professional talent
- Training for ITES human resource pool

5.11.3 The Cost Advantage

- A fully burdened cost of just $8 per hour when compared to the global average of $15
- Salaries—1/5th of the international average
- Operational costs less than 50% when compared to the other Indian cities
- Rentals lower by more than 60% in comparison to other Indian cities
- Power and water tariff among the lowest in the country

The IT policy gives considerable emphasis to the promotion of IT industry. It has been followed up by a number of activities, exclusively for the promotion of IT industry. The Technopark and the Software Technology Park in Thiruvananthapuram, IT parks promoted by KINFRA (Kerala Industrial Infrastructure Development Corporation) in Cochin and Calicut and Cochin Export Processing Zone are engaged in promotion of IT industries in the state. The Technopark in Thiruvananthapuram, with a built-up area of 800,000 ft^2 is India's most cost effective IT destination (Government of Kerala, 2001c). There are 54 IT companies in Technopark which provide employment for 5,000 professionals.

In IT hardware sector, the state continues its dismal performance. The data shows that the Technopark, that offers world class infrastructure suitable for both software and hardware companies, has only one unit for hardware production as against 53 software units. The domestic demand for PCs (personal computers) has not increased as expected. The quick estimate shows that total demand for PCs in Kerala was 150,000 in the year 2001. The local assemblers in the unorganized sector, catering to 60% of the market dominate the domestic PC market. The thriving business by the local assemblers is due to the low cost of the PCs offered by them. Many of these assemblers operate in grey market, offering attractive prices on account of tax evasion.

6. Analysis

The government has achieved success in creating a conducive atmosphere by removing the apprehensions of the trade unions, employees and the people regarding misconceptions about the computerization. In a democratic country like India, cooperation and participation of the people are necessary for dissemination and adoption of any new technology. The people-friendly projects like FRIENDS, SEVANA, PEARL, etc. have strong promotional effects and the general public has begun to realize the potential of applying IT in day-to-day life.

The pace of improvement that IT has brought in the registration department can be guessed from the fact that earlier it took two to three weeks for verifying the documents, but today it is done within less than an hour. In general, the perception of people has changed and they have begun to support and participate in IT promotion. The computerization in Kumarakom panchayat is a good example of the encouragement and participation of the people. "IT Kerala 2000", an exhibition cum investor meet conducted from 23–26 November 2000, the first ever national level IT event in Kerala, attracted nearly 30,000 business visitors and over 100,000 general visitors. This is testimony to the changing mood of the investors and the general public in Kerala (Kunnappally, 2001). The use of IT in enhancing the delivery of government services is aimed at creating a very responsive and transparent administration. This facilitates the empowerment of the people and satisfies their right to information.

The IT policy of the state aims to achieve everything possible with the application of IT, on a par with the developed nations. This seems to be a little over-ambitious. Does the state have the resources to implement all these projects in one stretch? Even though 3% of the state budget has been earmarked for the IT sector, this was not enough to implement all the projects. In order to ensure that benefits reach the rural poor, these projects are to be implemented throughout the state. This will involve a huge outlay of funds and the state should have a more realistic approach for fund mobilization. Moreover, the policy objectives have not been linked with the financial capacity of the state. Hence, the various projects could not be completed within the stipulated time frame. The financial limitation of the government has affected the pace of implementation of various E-governance projects. As suggested by Vittal (1999), the government should lease computers instead of purchasing them. If the government goes for leasing with a capital of Rs.500 million, Rs.5,000 million can be leveraged.

It can be seen from Table 15.1 that the developed countries have a high rate of PC penetration (World Bank, 2002). One of the targets of IT policy 1998 was to en-hance PC penetration to 1% of the population in Kerala by the year 2001. This itself was a modest target. However, IT policy 2001 does not target any figure, instead mentioning that an expert advisory group with leading experts from the Internet and computer industries has been formed under the IT department to recommend the enabling steps to accelerate the process of PC and Internet penetration.

Table 15.1
PC Penetration Rates in Selected Developed and Developing Countries

Country	No. of Computers per 1,000 Population
USA	585.2
Canada	390.2
Hong Kong	350.6
Japan	315.2
Germany	336.0
Norway	490.5
Switzerland	499.7
Finland	396.1
Singapore	483.1
UK	337.8
South Korea	237.9
Malaysia	103.1
China	15.9
India	4.5

Source: World Bank, 2002

However, the present low level of PC penetration in the state must be a matter of serious concern. If the state is to achieve success in E-governance as envisaged, it is necessary to have a higher level of PC penetration. There is a need for a clear strategy to enhance PC penetration. This is a pre-requisite for successful implementation of E-governance and improving the domestic market. The state should have a clear policy for encouraging individuals to buy PCs by offering incentives in the form of soft loans or tax deductions (Vital, 1999) . The policy of the Malaysian government can be followed in this case. The Malaysian government has launched an IT awareness campaign throughout the country and has provided tax deductions for first-time buyers of personal computers, among other things (Nain and Mustafa, 1998).

There is a distinct need to reorient the employees both in the government and private sector so as to meet the challenges of the future. Since the purpose of field level computerization is to improve management, it requires sustained training efforts and technical inputs. Training needs to be oriented towards the use of information by the workers, supervisors and managers for strengthening, planning and monitoring activities (Bhatnagar, 2000). Training can and must play an important role in improving the skills and quality of the government services. Employees should be encouraged to make learning a highly self-motivated activity for the acquisition of new knowledge and skills.

As it will be difficult to train a whole generation of government servants in IT in a short span of time, the private sector computer companies must be called upon to provide the equipment as well as services. This could be on a lease basis so that the government can meet the fund requirement. The problem of disposal of old equipment will also not arise. The government can thus ensure that it always has the latest IT system. Training programmes will have to be an ongoing feature in government, especially in the light of rapid changes in technologies and applications.

The success of the E-governance project depends on the E-awareness of the people. With the penetration of IT in all spheres of life it has become an important element in education. Hence, a comprehensive HRD programme is to be planned and implemented. Ongoing programmes of computerization in government departments, local self-governments, and other government and private institutions in Kerala will create a huge domestic market. The software companies engaged in software production for domestic sales are likely to achieve a quantum jump in their turnover. Consequently, there will be a large requirement for qualified software personnel. At present, the state is producing about 10,000 professionals annually with qualifications like, M.Tech, B.Tech, MCA (Master of Computer Applications) and PG Diploma in computer related subjects.

The state government has recently embarked upon a major policy initiative of liberalising the professional and higher education sector. The government has multiplied the number of technical seats in the IT sector (including engineering degrees and MCA). In this context, it should be admitted that the number of IT professionals that the state produces is much smaller than Karnataka, Tamilnadu and Maharashtra, which have been following a liberalized educational policy for some time. With the liberalization of the technical education sector in Kerala, the number of seats available is expected to be increased substantially. Through the liberalization of the technical education, the state is determined to ensure adequate manpower with appropriate skills in IT. As per the survey, conducted by the government of Kerala, there are 1,553 educational institutions engaged in computer education and training (Government of Kerala, 2000). It is necessary to ensure quality of the training courses in these institutions.

The IT policy proposes the establishment of Internet kiosks in every panchayat ward, accessible to any member of the public. The government should learn from the bitter experience of TV kiosks started in rural areas in the past that ultimately stopped functioning due to the lack of maintenance support. Another reason for the failure of TV kiosks was that the users did not find any added value, as the system was not interactive. Hence, the project on Internet kiosks is to be implemented with people's participation and it should also be capable of providing various services to the general public using the strength of IT.

The foregoing discussion suggests that as far as E governance is concerned, with a few exceptions, the state has achieved reasonable success. The state has initiated many innovative projects in E-governance. However, the pace of various E-governance projects needs to be accelerated to deliver services to citizens. It is a logical conclusion that all the government departments and local self-governments in the state will have excess manpower after successful implementation of various E-governance projects. This is likely to affect the availability of new job opportunities in the government sector. However, the growth of IT industry will create large-scale employment opportunities in the private sector, especially in IT enabled services, IT training and equipment servicing.

It is true that the state has taken a number of initiatives in tune with the IT policy. How far have these initiatives helped the state to realize the policy mission? Projects like FRIENDS, SEVANA, PEARL, and Information Kiosks are good examples for the use of IT for the benefit of the general public. However, these projects, which improve the service delivery of the government to the citizens, have been established in only a few locations, primarily in urban/semi urban areas. The majority of the rural population has yet to benefit from these projects.

In spite of the fact that the state has invested heavily, it could achieve software exports of only Rs.1,100 million (US$24 million) for the year 2000–01. It may be noted that during the year 2000–01, software export figures of neighbouring states Karnataka, Tamil Nadu and Andhra Pradesh were Rs.74,750 million (US$1,635 million), Rs.31,160 million (US$681 million) and Rs.19,170 million (US$419 million) respectively. According to the latest findings by the Electronics and Computer Software export promotion council (ESC), the state of Karnataka is the largest contributor to the export of computer software. The leading industrial state, Maharashtra, with its export contribution of Rs.42,750 million (US$933 million) is in second position. The state of Kerala has some consolation, when its performance is compared with the states like Gujarat and Punjab that have relatively higher per capita income.

In spite of the investor friendly policies of the respective governments, Gujarat and Punjab could achieve software exports of only Rs.1,080 million (US$24 million) and Rs.520 million (US$11 million) respectively.

Though there have been large variations in software exports from state to state, at the national level, the exports from Indian software industry continue to show impressive growth rates. In terms of Indian rupees, the C.A.G.R. (Compounded Annual Growth Rate) over the past five years has been as high as 62.3%. The industry exported software and services worth Rs.300 million in 1985. However, in 2000–01 a total export of Rs.283,500 million (US$6.2 billion) was achieved. (NASSCOM, 2001) (See Figure 15.1).

Figure 15.1
IT Software Exports in India, 1995–2001 (NASSCOM, 2001)

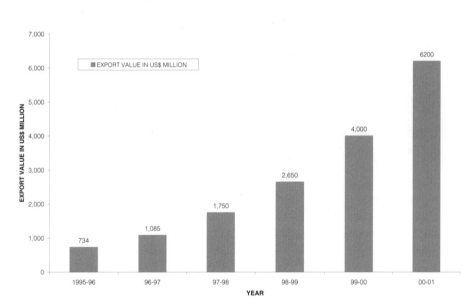

As per the survey conducted by the NASSCOM (National Association of Software and Services Companies), more than 3000 companies in India are engaged in the business of software exports (see Table 15.2). The top twenty-five software exporters (in the order of revenue) accounted for almost 60% of the software exports revenues in 2000–01 (NASSCOM, 2002).

The infrastructure projects undertaken by the government have been successful in creating a conducive environment for the development of IT in the state. As envisaged in the IT policy, the state has been developed as a fertile location for the growth of IT industry, through many infrastructure projects and other initiatives. The massive investment in infrastructure development will be justified, however, only when the state shows much improved performance in the development of its IT industry. The policies of the government and other initiatives have accelerated the development of human resources for IT. Through the liberalization of the technical education, the state is determined to ensure adequate manpower with appropriate skills in IT. The computerization of local self-government has been progressing at a fast pace. The various services from these offices like issue of certificates, licenses, tax collection, etc. are expected to become a reality in the near future and the rural population is also being brought under the ambit of application of IT.

Table 15.2
The Top Twenty Software Exporters
(Based on Revenue) in India (NASSCOM, 2001)

Company Name	Exports in Rs. Million
Tata Consultancy Services	28,702.60
Infosys Technologies Ltd.	18,529.40
Wipro Technologies	17,563.90
HCL Technologies Ltd.	11,269.20
Sathyam Computer Services Ltd.	12,412.20
Cognizant Technology Solutions	7,030.80
Silverline Technologies Ltd.	6,474.40
NIIT Ltd.	5,700.20
Penta Soft Technologies Ltd.	5,555.00
Penta Media Graphics Ltd.	5,480.20
Patni Computer System Ltd.	5,156.00
IBM Global Services Ltd.	5,060.20
Mahindra British Telecom Ltd.	4,499.80
HCL Perot System	4,392.40
DSO Software Ltd.	4,381.70
Mascot System Ltd.	3,398.70
Mascom Global Ltd.	3,386.90
i- Flex Solutions Ltd.	2,937.40
Tata Infotech Ltd.	2,880.30
Mphasis BFL Ltd.	2,833.10
Total	Rs. 157,644 million

7. Conclusion

The overall achievement in the IT sector is significant for a state that has only a meagre per capita income. Despite the financial constraints, the state has made significant achievements in E-governance, which shows the commitment of the government to enhance the delivery of services to citizens. However, it may be noted that a huge financial outlay is required to be mobilized for the fully-fledged implementation and maintenance of various E-governance projects throughout the state. In a democratic country like India, where 40% of the people are still living below the poverty line,

mobilization of the huge investment required for the implementation and maintenance of E-governance projects is not very easy as the state has a very high gross fiscal and revenue deficit. While all these government services are generally free or subsidized, the required resources should come from the returns of the IT industries.
Hence, promotion of IT industry through foreign and domestic investment needs to
be the priority of the government. The state should take a tip or two from the
neighbouring states of Tamil Nadu, Karnataka and Andhra Pradesh that have
achieved impressive growth of the IT industry.

Though the state has a large number of talented IT professionals and other intrinsic advantages, which are conducive to quick growth of the IT industry, the actual
performance is far below the desired level. Even though the state has established
world class facilities through exclusive IT parks and other IT infrastructure facilities,
the much-awaited investments from foreign and domestic promoters are not forthcoming to the desired extent and hence the facilities are being under-utilized. The
failure of the state to attract the leading Indian entrepreneurs and multinational companies should be studied in a broader perspective.

The state's poor performance in the IT industry sector can be viewed in the context of the overall industrial backwardness of the state. Nevertheless, the potential of
the IT industry for the future development of the state should not be underestimated.
For exploiting the advantages, the state should come out with suitable strategies to
encourage domestic and foreign promoters to invest in this ideal IT destination. In
this era of IT revolution, the state should not fail to exploit the opportunities for
achieving growth of IT industry.

References

Avgerou, C. (2000) Recognizing Alternative Rationalities in the Development of Information Systems,
The Electronic Journal on Information Systems in Developing Countries, 3, 7, 1–15, www.ejisdc.org.

Bhatnagar, S. (2000) Social Implications of Information Communication Technology in Developing
Countries: Lessons from Asian Success Stories, *The Electronic Journal on Information Systems in
Developing Countries*, 1, 4, 1–10, 2000, www.ejisdc.org.

Government of Kerala (1998a) *Information Technology Policy Document 1998*. Department of
Information Technology, Thiruvananthapuram.

Government of Kerala (1998b) *Report of the Advisory Council for Application of Information
Technology in Government*. Thiruvananthapuram.

Government of Kerala (1998c) *Report of the Task Force on IT Implementation in Government*. State
Planning Board, Thiruvananthapuram.

Government of Kerala (2000) *Report on Survey of Computer Institutions*. Department of Statistics,
Government of Kerala, Thiruvananthapuram.

Government of Kerala (2001a) *Census Report*. Department of Economics and Statistics, Government of
Kerala.

Government of Kerala (2001b) *Information Technology Policy Document 2001*. Department of Information Technology, Thiruvananthapuram.

Government of Kerala (2001c) *Economic Review-2001*. State Planning Board, Government of Kerala , Thiruvananthapuram.

Government of Kerala (2002) Advantage Kerala. Information Technology Mission, Trivandrum, www.itmission.org/content/AdvantageKerala.

Kumar, A (2001) Bridging the Digital Divide—Some Efforts from Kerala, Proceedings of the International Conference on Information Technology, Communications and Development, November, 29–31, www.itcd.net

Kunnappally, J. M. (2001) IT Kerala 2000: How Best to Conduct a B2B Meet, *Destination Kerala*, 4, 9, 8–9.

Nain, Z. and Mustafa, K. A. (1998) IT Strategies in Malaysia: The Multimedia Super Corridor, UNRISD Conference on Information Technologies and Social Development, June 22–23.

NASSCOM (2001) www.nasscom.org/it_industry/indic_statistics.asp

NASSCOM (2002) Indian IT Industry Statistics (2001) www.nasscom.org

National Geographic Traveller (1999) God's Own Country, *National Geographic Traveller*, Special Collectors Issue, October.

Vittal, N. (1999) Information Technology for the People, National Seminar I.T. Kerala, May 31.

World Bank (2002) *World Development Indicators*. Washington D.C.

About the Authors

Dr. K.G.K Nair obtained Bachelor degree in Mechanical Engineering from College of Engineering, Trivandrum, Kerala, India and Master Degree from Indian Institute of Technology, Delhi. He also obtained M.B.A and Doctorate subsequently. He has worked in various capacities like lecturer, Assistant Professor, Professor, Dean (P.G. Studies) and Principal in the department of Technical Education, Government of Kerala. He is presently working as Professor, Department of Business Administration, College of Engineering, Trivandrum. He has published several research papers and presented papers in national and international conferences. He has held various academic positions in University as member of faculty, Board of Studies. He is also a member of Professional bodies like Institution of Industrial Engineering, Institution of Engineers (India), and Indian Institute of Material Management.

Dr. P. N. Prasad obtained Bachelor degree in Electronics and Communication engineering from College of engineering, Trivandrum, Kerala, India. He obtained Master degree in Business Administration from Cochin University, Kerala and Doctorate in Management from Kerala University, India. He has worked in various middle and senior level management positions in KELTRON, one of the premier state electronics development corporations in India. Presently, he is working as a faculty member in State Institute of Rural Development, Kerala. He has published many research reports and papers on Information Technology, Rural Development and Marketing in national and international journals. He is also a visiting faculty in various management institutions.